NATURE'S
Classroom

A Program Guide
for Camps and Schools

Storer Camps

American Camping Association
5000 State Road 67 North
Martinsville, IN 46151-7902

ISBN #0-87603-110-6

Edited by
Jennifer R. Cassens
Cover Design by
Tom Dougherty

CONTENTS

PREFACE

This program and curriculum guide is simply a ''guide,'' not a pre-packaged outdoor education program. While some classes may be taught directly from this book, any class can be immeasurably improved by a teacher's research, reading, and creative thinking. Every class in this book evolved through research, experimentation, practice, ''borrowing'' of ideas from other centers, and responses from students and teachers. They have also been taught in various ways, touching upon different goals with different activities. The classes herein are not in final form which can never change; instead, they are only captured in these pages at a point in their evolution.

Every time a class is presented to a group of students, the instructor leaves his own indelible mark on them. Therefore, it is important that every instructor invests time and energy creating his own class, using this guide merely as a springboard. My desire is for this guide to be a seed from which ideas can grow and blossom as we all seek to improve education in the outdoors for our young people.

Regina D. Whitehead

ACKNOWLEDGEMENTS

This program and curriculum guide is dedicated to the many staff members of YMCA Storer Camps Outdoor Education Center who have expended their time, efforts, and emotions in a continual quest to develop a quality program since its inception in 1968. A special thanks goes to those who struggled through the early years by designing a program and curriculum from merely their creative imaginations. Foremost is Greg McKee who directed the program for eleven of those years. Special recognition also belongs to those who envisioned, researched, and wrote the first curriculum guide, which was written in 1978. These people were John Duckworth, Michael Fultz, Cindy Glick, Dave Glick, Warren Gartner, Debbie Glowik, Heidi Hargesheirmer, Greg McKee, Mary Mennel, Tim Milbern, George Notary, and Bill Sonnet.

The curriculum of a good program should be dynamic and vibrant. Because of an enthusiastic, creative Storer staff, who constantly search for new material and methods, more advances will have been made by the time this book goes to print. Thus, thanks also goes to those presently involved in experimenting with new ideas for the Outdoor Education Program, working with kids on a daily basis, hour after hour, in the outdoors.

Furthermore, this guide is dedicated to staff yet to come, who have a vision of the future in mind and will leave their signature on the classes, programs, and students of Storer Camps Outdoor Education Program.

Finally, appreciation is extended to Doris Musser who typed the manuscript for this publication.

Program Goals

Before embarking on a week of outdoor education, teachers should outline goals of the upcoming experience. While different schools have varying reasons for involvement in outdoor education, they all have some particular rationale for attending. To convince parents and administrators of the experience's validity and to get the full value of the outdoor education week, teachers should ask: "What do we want students to have within them in the end that they did not have at the beginning?" Once this question is answered, the planning process becomes much easier.

Outdoor education is a rather recent phenomenon, starting in Battle Creek, Michigan, during the 1920s. Its popularity, though, has soared only in recent years; and outdoor education centers have sprung up from California to Maine. Outdoor Education is simply teaching in the outdoors those things that cannot be taught as effectively in a classroom setting. If, for example, a math class is taught the same way in an outdoor setting that it is in a classroom, there is no purpose to move outdoors. The outdoor environment, therefore, must be used constructively for outdoor education to be valuable.

It must be stressed that outdoor education is an educational experience having important implications for students both in the outdoors and in the classroom. If the educational merit of the program is not emphasized, some parents might feel the program is a waste of valuable school time.

The following are some goals of an outdoor education program, which give teachers who are involved in planning a place to start:

OBJECTIVES:

The purpose of the outdoor education experience is:

1. To help students understand and experience the relationship between themselves and the natural environment.
2. To strengthen social relationships between teachers and students, students, and schools from different geographic and social backgrounds.
3. To help students gain a sense of independence and self-identity.
4. To give students a chance to succeed in a non-graded environment.
5. To help students become more aware of their senses.
6. To foster a spirit and attitude of inquiry in students.
7. To encourage students to actively participate in outdoor activities.

Housing and Cabin Leadership

Another important concern of teachers is student housing and leadership for students within those housing units. If cabins with a capacity of eight to ten students are used, one of the best methods for assigning students to cabins is to have each student choose a "'buddy'', and to put four to five of these pairs in a cabin. This allows each student to have a friend in the same cabin which helps to prevent some homesickness; yet, it allows teachers the final decisions in housing matters.

Each housing unit must have some adult leadership. The type of leaders chosen is crucial because their maturity, enthusiasm, and commitment make the week successful for the individual student. It is usually the school's responsibility to obtain housing leadership.

There are four sources of cabin leadership:

1. High school students, preferably seniors, may be selected by the school from its own district. These students should be carefully screened and thoroughly oriented. While high school students usually have a large amount of enthusiasm and commitment as leaders, they occasionally lack the good judgment necessary for taking care of a group of students. It is recommended, therefore, that an adult stay in the boys' and girls' cabin areas to assist high school students with their jobs. It should be kept in mind that this is also an educational experience for high school students; therefore, they should be made to feel like part of the experience. Feeling it is a valuable educational tool in the leadership training area, several school districts encourage high school students to become leaders.
2. College students may be acquired from colleges in the area which allow students to serve as leaders as part of an education or recreation program. College students, as a whole, are better leaders than high school students, but are harder to obtain due to their classes and assignments.
3. Teachers, student teachers, and teacher's aides may be possible leaders, although serving as cabin leaders places restrictions and extra responsibilities upon them. Some school districts feel it crucial that teachers stay in the cabins with students. They feel that the more teachers get involved in the week's activities, the more they and students benefit from the experience both at the outdoor education center and back in the classroom.
4. Parents of students, or other adults within the community, may provide the needed leadership.

Sample Outdoor Education Class

		A	B	C	D	E
MON.	2:00	Eye Opener	Eye Opener	Eye Opener	Eye Opener	Eye Opener
	3:15	Horse Sense	Enchanted Forest	Ecosystems	Tree Encounter	Pond Life
TUES.	9:30	Survival	Horse Sense	Tree Encounter	Pioneer Tools	Enchanted Forest
	10:45		Pond Life	Pioneer Tools	Ecosystems	Tree Encounter
	2:00	Horse Ride	Tree Encounter	Enchanted Forest	Horse Sense	Survival
	3:15	Incredible Journey	Horse Ride	Horse Sense Horse Ride	Pond Life	
WED.	9:30	Enchanted Forest	Pioneer Tools	Survival	Native American Life Fair	
	10:45	Pioneer Tools	Ecosystems			
	2:00	Native American Life Fair			Horse Ride	Horse Sense
	3:15				Enchanted Forest	Horse Ride
THURS.	9:30	Ecosystems	Survival	Activity	Enchanted Forest	Incredible Journey
	10:45	Tree Encounter		Incredible Journey	Survival	Pioneer Tools
	12:00	Cookout				
	2:00	Pioneer Tools	Incredible Journey	Pond Life	Survival	Ecosystems

Figure 1.1

It is crucial that leaders have some kind of orientation before the program begins, so that the student's experience runs smoothly and successfully. This may be accomplished in a combination of several ways:

1. A leader handbook, made available to leaders before their arrival at the center, may be used in orienting them prior to the outdoor education experience.
2. The outdoor education director or staff may meet at the high school or college for an orientation meeting.
3. The evening before students arrive, leaders may meet at the outdoor education center for orientation and a chance to become familiar with the facility.

At any rate, whatever method is used, the better the orientation process, the more smoothly the entire program proceeds.

Class Schedules

One of the great advantages of outdoor education is the reduced size of the class groups. If a center has cabins with a capacity for ten students, it may be possible to combine one boys' cabin group and one girls' cabin group to form a class or traveling group of 16-20 students. This method is effective for several reasons:

1. Persons in charge of housing units can be responsible for getting students to class.
2. Leaders can assist in teaching classes which could be structured to allow for their help. As a result, the number of students per leader is reduced to five or six to one, thus providing more opportunity for activity.
3. Students find it easier to follow schedules applying to housing units as opposed to individuals.

The schedule shown in Figure 1.1 shows how a group of 80-100 students (50 boys and girls) might be divided into five traveling groups (A, B, C, D, and E) and assigned to classes throughout the week. The class names would differ, of course, depending on the desires of teachers. Each class is taught five times during the week, once to each class group. Classes can be offered in either one- or two-hour blocks.

Figure 1.2 shows a suggested schedule for a week's experience in outdoor education.

Curriculum

Since the curriculum is the heart of the outdoor education experience, both time and care should be devoted to its formation. When selecting classes and activities for the program, the importance of using the outdoor environment should be kept in mind; and constant reference should be made to the objectives of the program.

Although a resident staff may be available at a center and is prepared to teach the entire curriculum, it must be emphasized that the outdoor education program should fit the objectives and goals of the particular classroom teachers. It should be a teacher's program; therefore, teachers must prepare students and carry the benefits of the outdoor education experience back to the classroom. Teachers' involvement throughout the week should be encouraged, but the decisions of when and how much should be entirely the teachers'.

Classes

As an EYE-OPENER, or introduction to the program and outdoor facility, take an orientation hike. Use this time in an attempt to open students' eyes to the educational nature of the outdoor experience. Games, explanations, detailed descriptions of the weather station, waterfront, horse barn or other facilities, questions to get students thinking, and nature tidbits — all can be used during this hike around the outdoor education center.

The following are short descriptions of some classes which may be offered in an outdoor education program. These classes are detailed in the following chapters with actual lesson plans.

Critters (Chapter 2)

AQUATIC LIFE (1 hour) — Following a discussion on communities, aquatic animals, and comparisons of lakes and swamps, students collect samples from a lake or pond. Back in the classroom, they examine the specimens under a

Possible Schedule for a Week's Experience

Monday

10:15	Arrive
11:00	Orientation
12:30	Lunch
2:00-4:15	Classes
4:45	Group games
5:30	Dinner
7:15-9:00	Evening activity
10:00	Lights out

Tuesday

6:30	Wake up, clean cabins, bathrooms
8:00	Breakfast
9:30-11:45	Classes
12:00	Lunch
1:30-3:45	Classes
4:00	Afternoon activities, Cabin leader meeting
5:30	Dinner
7:15-9:00	Evening activity
10:00	Lights out

Wednesday

6:30	Wake up, clean cabins, bathrooms
8:00	Breakfast
9:30-11:45	Classes
12:00	Lunch
1:30-3:45	Classes
4:00	Afternoon activities, Cabin leader meeting
5:30	Dinner
7:15-9:00	Evening activity
10:00	Lights out

Thursday

6:30	Wake up, clean cabins, bathrooms
8:00	Breakfast
9:30-11:45	Classes
12:00	Cookout
2:00	Class
3:15	Afternoon activites, Cabin leader meeting
4:15	Rest period
5:30	Dinner
7:30-9:00	Evening activity
10:00	Lights out

Friday

7:00	Wake up, clean cabins, pack clothes
8:00	Breakfast and evaluate week
9:00	Clean-up projects
11:30	Depart

Figure 1.2

microscope. Most commonly found aquatic animals can be visually charted for student discovery and identification.

CREATURE FEATURE (1 hour) — This class presents a hands-on approach to adaptations in mammals. After a short discussion of how animals must adapt, students examine mammal skins for adaptations which enabled the mammals to survive. Afterward, an adaptation auction, in which each mammal is described by three adaptations, is conducted. Using gold nuggets to purchase the animal, students bid on what animal they think is it. Finally, students attempt to match skulls and skins by examining eye placement, size, and teeth. This class makes an excellent winter choice since it occurs completely inside.

HABITAT HUNT (1 hour) — After discussion of habitats and animals signs, students go out to locate and identify animal signs in various natural areas. Upon their return, students show what they found, and the group interprets the information. Concepts such as the food chain, adaptation, and interdependence are discussed. The class encourages students to explore, discover, and interpret findings as they examine which animals live in certain types of habitats.

ECO-SYSTEMS (1 hour) — Oriented to introducing concepts of food chains and the interdependence of all living things, the class is made up of a game series. Activities include a predator/prey game, making a recipe for an ideal forest, a food web game, and several other simulation games. Students can view dramatically the dependence of an animal at the top of the food pyramid on the lower-level animals.

WILDLIFE MANAGEMENT (1 hour) — This is an introduction to the wildlife manager's job: his struggle with limiting factors, the obstacles he must overcome to control them, and the results to wildlife when he fails. Three simulation games dramatically show the effects as some students become predators and others prey. Limiting factors are added and taken away by the instructor, and the struggle for survival becomes intense. Then, through a unique discussion technique, a debate is held concerning the role of hunting and trapping as a tool for wildlife management. (Recommended for junior high school students.)

IN COLD BLOOD (1 hour) — This class represents a hands-on approach to adaptations of reptiles and amphibians. After a brief discussion on its history, students close their eyes and hold out their hands while an animal is brushed against their fingers. Once this is done, the animal is described and discussed; then everyone is given a chance to hold the animal. This process continues until all the critters are handled and discussed. The last few minutes of class are reserved for students to hold any or all of the animals.

SQUIRRELYMPICS (1 hour) — Students identify and discuss different species of squirrels and the adaptations and characteristics of each type. A large portion of class is held outdoors because students take part in activities designed to give them a feel for the skills that a squirrel must develop in order to survive. At the end of class, students try to do tasks that are easily completed by squirrels: tree clinging, log hopping, four-legged log run, and eating a peanut squirrel style. This class is best suited for fall or early winter.

HORSE SENSE (1 hour) — This class should be a prerequisite for the horse ride. In addition to information concerning the mechanics of the ride, safety around horses, and barnyard rules, the class also presents a fascinating look at the adaptations of the horse: its evolution through the years and its means of survival and protection. The class enables many students to overcome their initial fears of the horse, becoming comfortable in this awesome animal's presence.

HORSE RIDE (1 hour) — After a quick review of safety procedures and control of the horse, students are mounted on horses and proceed on a 45-minute trail ride through scenic woods and fields. Walking and trotting the horses, students can feel the pleasure of controlling their own animals under cautious supervision of trained horse staff.

Outdoor Feelings (Chapter 3)

DISCOVERY HIKE (2 hours) — Designed as an interpretive and recreational hike of about two miles, the hike can be designed specifically for the outdoor center area. It can be a fascinating hike, stressing geology, animal signs and habitats, plants and trees, and ecological concerns along the way. Games, activities, and interesting phenomena capture students' interests the whole way.

ENCHANTED FOREST (1 hour) — Students, having a chance to use all five, discuss the five senses and their importance. Afterward, while holding onto a rope, they are led on a blindfold hike, taking them through the enchanted forest of their imaginations. They stop along the way to feel, smell, listen, and eat various things to experience numerous sensations without the benefit of sight. Following the hike, they are asked to write about or discuss their sightless experience. This class is designed as a sensitivity class, helping students become more aware of their surroundings.

CREATIVE EXPRESSION (1 hour) — This class encourages students to observe and experience the world around them and to describe it in their own words. They have a chance to do some drawing and creative writing and to share their thoughts with others.

FOREST MANOR (1 hour) — In an indoor winter class/game an eight-story condominium is proposed to be built in a wilderness area at the center. Students are given role cards (soil, unemployed worker, mayor, farmer, rabbit, etc.) and asked to vote in favor or in opposition of the proposal. After a mock town meeting is held, more real-life situations are discussed to point out environmental issues in construction and development.

Trees and Such (Chapter 4)

TREEE-MENDOUS (1 hour) — Students gather leaves, examining them closely, and then learn how to use a leaf key to identify the individual leaves. The important concept is not the memorization of leaves, but the process of gathering information, interpreting data, and thinking deductively.

FORESTRY (1 hour) — Students examine a stand of white pines and discover that the stand needs thinning. They then work as foresters, using forester's tools to cruise and measure the timber. Basic math skills are used to figure basal area; and, from this, they decide how many trees must be removed. With the information presented in the introduction of the class, they mark the immature and dying trees for thinning.

INCREDIBLE EDIBLES (1 hour) — Usually offered in the late spring or early fall, this class consists of a hike which introduces students to the edible plants of an area. They learn which plants should not be touched, how edible plants can be cooked, and how Indians and pioneers used various plants.

SUCCESSION STUDY (1 hour) — Students learn concepts of succession by studying four different areas of land leading from the lake toward the oak forest. Emphasis is placed on scientific method, small group work, student involvement, and gathering and interpreting data. Examples of succession, which might be observed in their own yards, are discussed, and further ecological results are considered. This class is effective only if there is access to a lakeshore bordered by a forest with various stages of succession.

Back in Time (Chapter 5)

PIONEER CRAFT FAIR (2 hours) — This activity class shows the daily life of a frontier pioneer. The first hour is spent demonstrating four basic crafts: spinning, candlemaking, soapmaking and butter churning. The second hour is a fair in which students can select as many activities as they desire to do. With the teachers' and cabin leaders' help, the staff can offer candlemaking, cooking in a Dutch oven, cornhusk dolls, corncob pipes, and many more pioneer crafts. This is an exciting class which gives students a real feel for the tasks their forefathers had to do to exist on the frontier.

NATIVE AMERICAN LIFE FAIR (2 hours) —
Students are divided into four groups; after an introduction, they become Native Americans of the area. Rotating through a series of activities, they may play war and gambling games, set traps and snares, do some cooking of fry bread and parched corn, learn Native American sign language, and learn how to use and make weapons. The cultimating activity is a large council meeting in which a Native American agent tries to convince them to turn their land over to the U.S. Government. Students play the roles of chief, medicine man, scouts, and others as they tune into how the Native Americans felt during these times.

PIONEER TOOLS (1 hour) — Students discuss early settlers of the area: when and why they came and what they brought. Students then learn the step-by-step process by which these settlers built their barns, watching the actual tools being used. The last part of class allows them to handle and use the very tools pioneers used over a hundred years ago. Students can get the feel of working with wood as they make shingles, mauls, and stools with tools such as froes, adzes, broad axes, and draw knifes.

MICHIGAN COUNTRY (2 hours) — Students learn the basic concepts of surveying and plotting through a simulation activity which depicts how Michigan was settled in the early 1800s. They group into families and role play as early settlers, attempting to establish their homesteads. Students choose their family's income as they scratch out an existence by trapping, trading, farming, and dealing with the area highwaymen. Group interaction is a vital element as students gather for town meetings to discuss their problems and frustrations in the new settlement. Historical details can be adapted for the specific area of an outdoor education center. (Recommended for junior high school students.)

HOMESTEADING (1 hour) — This class gives students a chance to role-play some early Americans traveling westward. They make decisions as families: why they are moving, what they need to take, and what they will do when they arrive. They have a chance to experience some of the difficulties pioneers faced.

Challenge (Chapter 6)

ORIENTEERING (1 hour) — This class teaches the basics of compass use. It involves converting meters to paces, direction finding, and working in compass teams, and prepares students for using the compass in several other classes.

INCREDIBLE JOURNEY (1 hour) — This is a combination orienteering and obstacle course. At the end of each compass reading is an obstacle with the group must overcome. Cooperation is essential and overcoming the obstacles develops group cohesiveness. Obstacles include getting the group over a poisonous vine, down a path after each member is blinded, and crossing over five stumps to ford a river.

MISSION IMPOSSIBLE (2 hours) — This is a challenging cooperation course, requiring teamwork, patience, and thought to complete. Using a compass and covering long distances, students come upon numerous obstacles and challenges which their group needs to overcome. (Recommended for junior high students.)

SURVIVAL (2 hours) Students learn the basic survival skills: use of the compass, fire building, shelter building, signaling, and dealing with panic. Then they are sent on a simulated survival exercise involving application of these skills. A tremendous amount of group participation is needed to survive this simulated wilderness experience. Students actually build a fire, boil water, build a shelter, and signal for rescue.

WILDERNESS RESCUE (2 hours) — Students spend the first hour discussing wilderness survival situations: hypothermia, general first aid, water rescue, signaling, fire building, and search rescue. The second hour they are involved in an intense simulation of six emergency situations where they successfully try to rescue their leaders, earning point values as well as learning cooperative skills. (Recommended for junior high school students.)

THE BEAST (1 hour) — This class involves group cooperation and communication skills. Students are divided into teams of four; each then chooses the job of builder, observer, information relayer, or buyer. The goal is to build an object (the beast) exactly like the one designed, which has been hidden from everyone's view. The observer, the only one who can see the beast, reports what is seen to the relayer, who then reports to the buyer. The buyer then purchases what is needed to build the beast and reports to the builder, giving the builder instructions and material. The builder then tries to construct the beast. Following its construction, students discuss their accomplishments and frustrations. (Recommended for junior high school students.)

The Good Earth (Chapter 7)

FUTURE SCORE (1 hour) — Students identify and discuss current energy sources and how they effect the environment. From this list, students who have grouped themselves into families, choose and purchase an available energy source to supply a dwelling they have created. At this point, the families embark on a twenty-year adventure. Dealing with bill collectors and natural hazards, their main goal is to keep their ideal dwelling intact until twenty years have passed. Factors that students find important when choosing their energy sources are availability, resource renewability, environmental impact, installation, and maintenance costs. Possible solutions to the impending energy crisis are discussed, and practical measures such as conservation, insulation, and dwelling design are emphasized. (Recommended for junior high school students.)

WEATHER MAPPING (1 hour) — Students use observation skills in reading simple weather instruments in a designated area which has varying physical features. They collect data at four different spots in an area and, with the use of an overlay map of the area, record their information and draw conclusions.

SOIL SURVEY (1 hour) — Students become familiar with various types of soil, layers of soil, and soil testing kits. They examine different soils and test them for texture, pH factor, permeability, and color. They discover which soil is best for which use.

MAPPING (1 hour) — Students learn the basics of mapmaking. They learn how to transfer a large area and objects included in it onto graph paper, making their own map of the area. This class involves measuring distances and using the compass.

JIGSAW (2 hours) — This is a culminating activity combining other classes such as Mapping, Habitat Hunt, Creative Expression, Weather Mapping, and others. Students divide into several groups and are given a plot of land to study. They work together to create a comprehensive study of this specific area.

Winter Watch (Chapter 8)

WINTER WATERLAND (1 hour) — This is a pond life class taught during winter, allowing students to discover the cause and effect of ice formation on a lake. With spud bars and plant hooks, students chip holes in the ice and collect plant samples from the bottom of a lake. Returning to a classroom, they discover life on plants while using microscopes to identify aquatic animals.

WINTER BIRDS (1 hour) — This class usually can be done from November through March in most areas. It begins with an introduction to winter birds and how they survive winter weather. With the use of bird study skins, students can better understand the reasons for different beak shapes, feet, wings, and feathers. After instruction on how to use binoculars, students make observations on winter birds at the bird blind. The class is designed to teach observation and appreciation of bird adaptations and habits.

TRACKING (1 hour) — This class is restricted to months when snow covers the ground. The first part of class is held indoors with discussion of habitat, hibernation, and animals that remain active throughout the winter. Animal signs such as tracks are illustrated. The second part is held in the natural setting. An activity, using scent tracking techniques with colorful, aromatic scent sprayed on the snow is conducted. The trail is followed by students, much like predators after their prey, to find the source of scents.

SKIING (1 hour) — This special outdoor activity class gives students a chance to explore an exciting sport. After some basic instruction concerning care of equipment, turns, stops, gliding and falling, students are taken over groomed ski trails through woods and fields. Students desiring to ski more should be able to have one more opportunity later in the week.

TWIG-O-MANIA (1 hour) — Students interview a tree, finding out all they can about it, and then learn how to use a tree key to identify the tree. The important concept is not the memorization of trees, but the process of gathering information, interpreting data, and thinking deductively.

Afternoon Activities

These activities are more recreational than the classes and allow students to enjoy the outdoors. Students meet at a central location and are divided into activity groups. Everyone must participate since student supervision at the outdoor center is important.

Activities may include:

- Animal tracking and plaster casting
- Broom ball
- Conservation projects
- Early American crafts (spinning, buttermaking, soapmaking, dyeing, candlemaking, cornbread making.)
- Field games
- Fishing
- Hikes to interesting areas
- Ice skating
- Native American games
- Indoor games and activities for bad weather
- Lumberjacking
- Nature craft projects
- Photo hikes
- Planning skits
- Scavenger hunts
- Setting up compass courses
- Singing
- Snow sculpture
- Swamp stomps

Evening Activities

These are some suggested evening activities which are recreational as well as educational. They serve to bring everyone together for enjoyable common experiences. The following activities have proved very successful:

Relay Games — Usually a good, first-night activity, relays are run among cabin groups. The purpose of relays is to bring cabin groups together, giving cabin leaders a chance to get to know their cabin groups and to develop cabin spirit. They can be structured so that everyone has a chance to succeed and an individual's athletic ability is of no particular benefit.

Campfire — This favorite activity of students includes singing, skits, and a story — inside or outside around a campfire.

Night Hikes — Designed to get students to feel comfortable in the outdoor world at night, this activity encourages them to use their senses of hearing, touch, and smell in discovering the excitement of that environment.

Square Dance — Many groups consider this a favorite activity. Students may be uneasy at first; but after seeing teachers, counselors, and staff having fun, there is no end to their enthusiasm.

Skit Night — Students present two- to five-minute skits or mini-plays to the assembled group. They should be about the week's activities, and should be carefully screened. Skits are a good way to bring the group together for an enjoyable activity.

Counselor Hunt — Staff, teachers, and cabin leaders go out to parts of camp and hide. Cabin groups try to find staff and have them sign a piece of paper. The group with the most signatures wins.

Medicine Man Scavenger Hunt — Each cabin group tries to find all the ingredients needed for the medicine man to save the chief's life. At the end of a given time period, groups assemble to see which group could save the chief's life. Items are all natural objects such as a gall, something never seen before, something that changes, an animal bone, some hair, a stick in the shape of an S, an oak leaf, three different kinds of seeds, or something an Indian could use.

ABC Hunt — Cabin groups go outdoors and try to find natural objects that begin with A, B, and so on through Z. The group with the most objects wins.

Ecodrama — A nature charade, this fairly short activity can precede night hikes. Students act out natural phenomena without props or sound effects. Examples which might be used include forest, tornado, snake eating a mouse, or a waterfall.

Native American Council Meeting — A reenactment of an 1800's council meeting, this evening activity is a good follow-up of the Native American Life Fair. Students role-play members of a local Native American tribe (hunters, scouts, elders, young warriors) who meet with a U.S. government agent to discuss a treaty signing that gives land to settlers. Much discussion is employed with the ultimate decision held in the hands of the chief. It is important to fit the activity as nearly as possible to actual historical events and people.

Dutch Auction — An activity designed to increase group cooperation and imagination, this activity requires cabin groups to collect various items such as toothbrushes, flashlights, bandannas, and stuffed animals from each member. An auctioneer calls for an item, and all cabins that present that item receive points. Several requests are not tangible things but acts the group must perform or use their imaginations to complete.

This is a good activity to get cabin groups working together.

Lorax — Dr. Seuss creates an excellent environmental story which can be acted out by staff with help from students. It is a good way to conclude a week and discuss ways of protecting our environment.

Native American Myths — Students hear a story about a Native American myth and divide into groups of five to seven persons. The groups are given a set of cards with people or items so they can make up their own Native American myths.

Outdoor Education Center Equipment

Below is a list of equipment which an outdoor education center may wish to have available for visiting schools:

Balls — volleyballs, basketballs, playground balls
Butter churn and soap crock
Cards, drop spindles, and spinning wheels for spinning wool
Compasses
Film projector
Fishing poles
Ice augers
Tipi
Hot plates
Microscopes and aquatic life equipment
Saws, hammers, and other tools
Soil test kits and equipment
Thermometers, wind-direction and wind-speed indicators
Weather station equipment (barometer, hygrometer, etc.)

Some craft supplies may be important: paper scissors, crayons, etc. Teachers, however, can be encouraged to bring this equipment with them since a supply may be difficult to maintain at an outdoor education center.

Pioneer Crafts Fair Equipment List
(Based on 100 Students)

A. BUTTER
   1. Butter churn and/or small jars with tight lids
   2. 12 half-pints whipping cream

B. SOAP
   1. 4 pounds lard
   2. 13-ounce jar lye

C. CORN NECKLACES
   1. Elastic or fishline
   2. Paper clips
   3. Corn

D. CORNHUSK DOLLS
   1. String
   2. Cornhusks

E. CANDLES
   1. 8 pounds wax
   2. 10 yards wick material
   3. Hot plate

F. CORNBREAD
   1. 1-pound bag cornmeal

G. DYEING
Nothing needed

H. CORNCOB PIPES
1. Corncobs
2. Knives

I. WEAVING GOD'S EYES
1. Poster board
2. Balls of yarn (various colors)

SPECIAL EQUIPMENT SCHOOLS SHOULD BRING

1. Any art supplies used during the week
2. Supplies for Early American Crafts Fair

Suggested Individual Student Clothing and Equipment List
(One Week's Experience)

BEDDING

Sleeping bag or 2-3 blankets and sheets.
Pillow and pillow case, if desired
(Be prepared for cool nights.)

CLOTHING

It is suggested that children bring old but clean clothing. New clothing may look like old clothing when children get home. Clothing should be marked with the student's name.

1. One pair of pajamas
2. Two pairs of shoes
3. Daily change of socks and underwear
4. Heavy and light shirts
5. Warm jacket and sweater
6. Handkerchiefs
7. Raincoat
8. Stocking hat (a must in winter)
9. Two or three pairs of trousers or jeans
10. Gloves or mittens and scarf (in cold weather)
11. Waterproof boots

TOILET ARTICLES

1. Toothpaste and brush
2. Soap/shampoo
3. Two bath towels and washcloth
4. Lip balm
5. Comb/brush

GENERAL EQUIPMENT

1. Flashlight
2. Stationery and stamps
3. Pencils and notebooks

OPTIONAL EQUIPMENT

1. Binoculars
2. Camera
3. Compass
4. Fishing pole
5. Ice skates (in season)

DO NOT BRING

1. Money
2. Radios
3. Knives
4. Food (includes soda pop and candy)
5. Chewing gum
6. Comic books and card games
7. Firearms and archery equipment
8. Axes and saws
9. Matches
10. Curling irons
11. Blow dryers
12. Electronic games

Suggested Policies

One objective of nearly any outdoor education center is facilitating a student's learning in the outdoor environment. To facilitate that learning, there are certain policies which may prove to be important.

The use or possession of alcoholic beverages and drugs should be prohibited. Smoking by staff may be permitted in a parking lot, teacher's quarters, or other designated areas, but may not be permitted on the grounds, in buildings, or around students.

In an effort to preserve the natural environment, cars may be limited to parking areas and designated driveways. Trailbikes or snowmobiles may be prohibited or limited to certain areas. Pets are a threat to the animals and their habitats and may not be permitted.

There may be a great deal of wilderness land surrounding a center; thus, students, unless accompanied by an adult, may be required to stay in designated areas. The center should not be responsible for students or persons from the school who leave the property during the educational experience.

The outdoor education director, nurse, or teachers are the only individuals who should administer first aid to students. A nurse may hold a sick call during specific times. Emergencies should be taken to a local physician or emergency room. A signed copy of a health form probably will be necessary. (See Figure 1.5.) The outdoor education director, nurse, and teachers should be notified immediately of any illness or injury. School systems, at their expense, should provide limited accident insurance coverage for participants, if such insurance is required.

In some states, it is a law that a child shall not be deprived of food, isolated, or subjected to corporal punishment or abusive physical exercise as a means of punishment. The center director should be consulted concerning the correctness of a certain disciplinary procedure if it is doubted.

A waterfront or swimming area should not be used without prior arrangement with the center and proper supervision.

Since one of the objectives of an outdoor education experience is to help students become more independent, they might be asked to take on certain responsibilities for the center and program. These might include giving weather reports, raising and lowering the flag, cleaning restroom facilities or cabins, setting tables, and sweeping floors. If staff and teachers show enthusiasm for these responsibilities, the students will respond positively to them.

Figure 1.3 is a list intended to help teachers visualize their week's program. There may be thirteen or fourteen class periods offered in a week, usually two classes in the morning and two in the afternoon, with classes only on Monday afternoon and none on Friday.

Teacher Curriculum Checklist

_____ Eye-Opener (1 hour)	_____ Pioneer Tools (1 hour)
_____ Aquatic Life (1 hour)	_____ Michigan Country (1 hour)
_____ Creature Feature (1 hour)	_____ Homesteading (1 hour)
_____ Habitat Hunt (1 hour)	_____ Orienteering (1 hour)
_____ Eco-systems (1 hour)	_____ Incredible Journey (1 hour)
_____ Wildlife Management (1 hour)	_____ Mission Impossible (2 hours)
_____ In Cold Blood (1 hour)	_____ Survival (2 hours)
_____ Squirrelympics (1 hour)	_____ Wilderness Rescue (2 hours)
_____ Horse Sense (1 hour)	_____ The Beast (1 hour)
_____ Horse Ride (1 hour)	_____ Future Score (1 hour)
_____ Discovery Hike (2 hours)	_____ Weather Mapping (1 hour)
_____ Enchanted Forest (1 hour)	_____ Soil Survey (1 hour)
_____ Creative Expression (1 hour)	_____ Mapping (1 hour)
_____ Forest Manor (1 hour)	_____ Jigsaw (1 hour)
_____ Tree-mendous (1 hour)	_____ Winter Waterland (1 hour)
_____ Forestry (1 hour)	_____ Winter Birds (1 hour)
_____ Incredible Edibles (1 hour)	_____ Tracking (1 hour)
_____ Succession Study (1 hour)	_____ Skiing (1 hour)
_____ Pioneer Crafts Fair (2 hours)	_____ Twig-o-mania (1 hour)
_____ Native American Life Fair (2 hours)	

Figure 1.3

Teacher Checklist

Task Completed

Outdoor education date confirmed: _____   _____

Money from students collected   _____

Housing roster collected   _____

Physical forms returned by students   _____

Master copy of physical problems made   _____

Buses arranged   _____

Housing leadership finalized   _____

Equipment and supplies gathered   _____

Schedule completed by teachers or outdoor education center director   _____
   NOTE: Schedule should be in the hands of all people involved (teachers, director, etc.) at least one week before the experience at the outdoor education center

DATES TO REMEMBER

Teacher meeting   _____

Parent meeting   _____

Student orientation session   _____

Follow-through program   _____

Figure 1.4

Outdoor Education Health Form

School _____

Student's Name _____ Age _____

Address _____ Birthdate _____

City _____ State _____ Zip _____ Home Phone _____

Father's Name _____ Work Phone _____

Mother's Name _____ Work Phone _____

Family Doctor _____ Doctor's Phone _____

Medical Insurance Company _____ Policy # _____

If parents not available in an emergency, notify:

Name _____ Phone _____ Relationship _____

General information necessary for your child's protection and care:

1. If your child must take any medication, send medicine in original container and label. Please give the following information:

   a. Name of medications: _____ Dosages: _____

   b. Times usually taken: _____

   c. Reason for taking the medications: _____
   _____

2. Allergies (food, insect bites, drugs, others): _____

3. Has your child been exposed to any communicable disease within the past ten days? _____

   If so, what disease? _____

   Are there any physical activities in which your child should not participate? _____
   _____

4. Date of last tetanus shot, if known: _____

5. Any other information we need to know about your child: _____
   _____

It is necessary that the school and camp authorities know your child's physical and mental condition. If you have any doubt that your child is in good health, have him or her checked by the family doctor and forward the report to school.

I hereby give permission for emergency treatment for my child in case of accident or illness, and for normal treatment by the nurse at the outdoor education center. In case of emergency, I may be reached at:

Address _____ Phone _____

Parent's Signature _____ Date _____

Figure 1.5

Critters

chapter 2

Aquatic Life in Pond and Swamp

Objectives:

1. To increase students skills of observation by looking for very small animals.
2. To increase students' awareness of the abundance and diversity of life in the lake or swamp.
3. To teach students an appreciation for the value of a natural resource, such as a swamp, which is usually regarded as useless.

Equipment:

1. Golden Guide Books, *Pond Life* (See chapter bibliography.)
2. Charts showing most of the common swamp animals
3. Eight microscopes, including slides and blister slides (Single-power 50 X Blister Microscopes work well.)
4. Eye droppers, kitchen basters, petri dishes or small cups
5. Collecting jars or cups and buckets
6. Aquariums containing fish from a lake and a swamp aquarium
7. Two dip nets and a seining net for lake studies
8. Additional equipment for studies of the lake such as plankton nets, a Hach kit (to do PTT and other chemical tests), and extra dip nets, if desired
9. A pond, swamp, or lake (Because the lake and swamp are frozen during the winter months, most of the animals are inactive and unavailable. Therefore, this class can be taught in the fall before the freeze and in the spring after the thaw.) (Also see the Winter Waterland class.)

I. Pre-Experience Discussion Ideas

(Page numbers refer to Reid, Zim and Fichter, *Pond Life* from the chapter bibliography.)

A. What is *ecology?* It is the study of plants and animals, their interrelationships and ways they fit into the environment. In this case, the pond or swamp is the environment.

B. Give an example of a *food chain.* A bass eats a blue gill, which ate a minnow, which ate a mosquito larva, which ate a plant. Each organism is one link of the chain. However, this is incomplete. Not only are blue gills food for bass, but also for pike. Likewise, blue gills feed on more than just minnows; they also feed on small sunfish, rock bass, and others. Now branch out from your original food chain by taking into account these other animals. You end with a web-like affect, or *food web.* (pp. 22-23)

C. How do ponds appear? They may be the result of glacial action, the sinking of ground to form a depression, or man's impact. What is happening through time? As plants and animals die, their bodies decay, settle to the bottom, and begin to fill the lake. What will happen in the future? The lake eventually fills to form a swamp, then a meadow; and when it becomes dry enough, trees finally grow. This series of events is called *succession.* (pp. 24-25) (For further information, see the Succession Study class.)

D. Animals in Different Pond Environments:

1. Surface — water striders, whirligig beetles, springtails, spiders
2. Open water — bass, sunfish, other large fish
3. Bottom — catfish, dragonfly larvae, snapping turtles, clams, crayfish
4. Among weeds — minnows, insect larvae, snails, tadpoles (pp. 17-21)

E. Use microscopes in the classroom so that children know how to use and care for them before coming to the center.

II. Introductory Discussion

A. Community

1. What is a community? In human terms, a community includes people not only living together, but interacting with each other. In nature, a community includes all plants and animals living together and coming in contact with each other in a similar environment, such as a pine forest, field, or pond.
2. What is included in a pond or swamp community? All the plants and animals as well as water, mud, and other substances are part of the community.

3. What would happen if all the bass, pike and other meat eaters were removed from a lake community? What would happen to the blue gill and crappie populations? The populations would overcrowd with no predators to control them, resulting in the consumption of all their available food. The final result would be mass starvation. (Also see the discussion of communities in the Habitat Hunt class.)

B. Comparison of Lake and Swamp

1. A lake is larger and deeper than a swamp.
2. A swamp has rooted plants growing throughout, while a lake only has them along the shore.
3. A swamp has more mud and decaying matter on the bottom, yellowish-brown water, and natural oil from continual decomposition found on the surface.
4. The water temperature changes more rapidly in the swamp because the water is shallower and the sun can completely penetrate it.
5. A swamp is stagnant; a lake is not. In lakes there are currents under the surface, waves caused by the wind, and other splashing around. The swamp, however, is not affected with these movements.
6. A swamp has no fish because they need a lot of free oxygen which is not found in the swamp due to lack of movement and the decaying process that consumes much of it.
7. More diversity and abundance of animal life are in the swamp because more food and shelter are available in the form of plants and debris. If all animals are removed from the swamp and put on one side of a balance scale and all the animals from an equal-sized area of lake are placed on the other side, which side is heavier? The swamp side is because ten two-foot-long bass may be in that area of the lake, but not much else. However, the swamp has hundreds of thousands of small and microscopic animals which, when put together, weigh much more than those fish.

III. Collecting

Because of the great abundance and diversity of life in a swamp, it is recommended that the class go there rather than to a lake. Have them form pairs. Give one collecting container (plastic party glass, baby food jar, etc.) to each pair. Before going out, discuss the following:

A. Where can the most animals be found in the swamp? They are usually among the vegetation, which offers food and shelter.
B. Do not step in the water. Why? Besides getting wet, you stir up mud and silt and scare away animals.
C. When you find an animal, scoop it up. After looking at it, pour the animal into one of the large containers (gallon jars or buckets) so it can be brought back

for examination under the microscopes.
D. If you do not see any animals, take a random scoop through the vegetation, let the water settle in the container, and look closely. Almost every time there are at least a few animals — perhaps five or six fresh-water shrimp or mayfly larvae, a dozen cyclops, or fifty to sixty seed shrimp.
E. Do not be afraid to stick your hands in the water because none of the animals sting or bite.

IV. Microscope Work

A. Find an animal in one of the jars that you want to observe. Suction the animal into a kitchen baster (large plastic dropper). Remember to squeeze the bubble of the baster before sticking it into the jar so it does not stir up the water.
B. Empty the baster into a shallow container.
C. Suction the specimen with a small eye dropper.
D. Empty the eye dropper onto a microscope slide.
E. Using your finger, wipe off some of the water.
F. Put the slide under the microscope and use the knobs to get the animal into sharp focus.
G. When you are finished looking, dip the slide into the water to let the animal go before it dries up and dies.

V. Alternative Activities

A. Use a plankton net to collect floating microscopic life from the lake.
B. Study the pH of the water. Compare the results of the lake and swamp.
C. Compare the temperature of the lake at different depths.

VI. Post-Experience Ideas

A. Go to a body of water near your school to conduct another aquatic class. Note the similarities and differences between that body of water and the swamp or lake at the center. What are the reasons for the differences?
B. Using the specimens the students collected, set up a swamp or lake aquarium in the classroom.
C. Raise tadpoles in the classroom. Note the changes in their development.
D. Collect micro-organisms without even going to a swamp. Just place some hay or grass in a jar of water. After leaving the jar in a dark room for about a week, examine the water. You will be amazed at the animal life that appears.
E. Talk about the wise use of water resources.
F. Discuss the value of a swamp. Not only does the swamp provide a home for animals ranging from microscopic organisms to pheasants and deer, but swamp animals, themselves, provide a valuable food source for many other animals.

Creature Feature

Objectives:

1. To help students recognize the physical characteristics of an animal as *adaptations*.
2. To help students relate adaptations to the animal's needs and environment.
3. To give students a chance to see and handle twenty-five common mammals found in the home area.

(Having students name and identify mammals at the end of class is not an objective. The main objectives deal more with students gaining familiarity with mammals and the fact that each has certain characteristics, enabling it to survive.)

Equipment:

1. Display shelf for mammals and skulls
2. Twenty-five or more mammal study skins and as many skulls as possible
3. Mammals poster (Pennsylvania Game Commission poster ''Mammals of the Farm'')
4. Six small containers
5. Enough kernels of corn (gold nuggets) for each group to have thirty

This class may be taught completely indoors. Great care must be taken to insure that students properly handle the skins. They must be instructed to gently pick them up —not by feet or tails — and not use them as toys. Students should be instructed to not pet the animals and only pick them up for examination. Animal skins should be kept dry and clean. Placing skins on carpeting is dangerous, since claws stick in the carpet and legs may be damaged when the skin is lifted.

I. Introduction

How many of you have changed schools before? If a teacher in the new school came to you and said, ''You are going to have to adapt to this new school,'' what would that mean? How would you have to adapt? You would have to adapt to new teachers, new kids, new books, and a new building. Do animals have to adapt to new books and buildings? No, but an animal might have to adapt to other things. Physical characteristics which help an animal adapt are called *adaptations*. What kinds of things might make it necessary for animals to adapt?

A. Environment

   1. What is a mole's environment? Underground.
   2. What adaptation does it have to help it live there? It has large, powerful feet for digging and moving through the soil, no eyes, and fur which can move in any direction.

B. Food

   1. Can you tell from this skull what kind of food this animal eats? Plants or meat? (Show a meat-eater's skull.)

   2. What adaptation does it have that enables it to eat meat? Long, sharp teeth.
   3. Where are its eyes located on the head? Why? They are on the front so it can see to find prey. Eyes in the front also make binocular vision possible so it can judge distances and see depth.
   4. What kind of food does this animal eat? (Show a plant-eater's skull.)
   5. What kind of teeth does it have compared to a meat-eater's? It has flat and broad teeth which are for chopping and grinding off plants.
   6. What kind of food do we eat? We eat meat and plants. Use your tongue to feel the inside of your mouth. What kind of teeth do we have? We have incisors and grinding teeth for both meat and plants.
   7. Do adaptations apply only to animals? What else makes adaptations necessary for animals?

C. Enemies

   1. Where are the eyes on a beaver or other plant eaters such as horses, rabbits, and squirrels? They are located on the side of the head.
   2. If the animal is bent over while eating, it can still see a fox, or another predator, sneaking up.

D. The Weather

   1. Some animals grow more hair in winter.
   2. Some migrate or hibernate.
   3. Some eat more food and store body fat.

E. All living things, plants and animals, have many adaptations which help them survive. We will look at a few mammals and try to discover some of their specific adaptations. What makes mammals different from other animals?

   1. They have live birth, not eggs.
   2. They nurse their babies.
   3. They have hair.
   4. They are warm-blooded.

II. Activities

A. Examining Study Skins

At this time, the instructor demonstrates the proper way to handle the mammal skins and impresses upon the group the importance of treating them with care. He also explains that the animals were found dead and were not trapped or shot specifically for the class, since this is often a concern of students. While students are arranging themselves into five groups, the instructor places about five skins at each station located around the room. When this is done, he gives the meaning of each mammal's tag and tells what is going to be done next. One side of the tag has the name of the animal; the other lists the main adaptations which characterize it. (See Figure 2.1.) Each group goes to one station and examines the skins for

Figure 2.1

adaptations. They try to decide where the animal lived, how it got food, how it escaped predators, and how it dealt with weather changes. The instructor has each group rotate every few minutes to the next station. At the end of ten or twelve minutes each student has had the opportunity to examine each skin.

Assistants, who may be either cabin leaders or teachers, can rotate around the room to remind students to look for certain characteristics in each animal. Students often get so engrossed in examining the skins that they forget what they are supposed to do. The instructor and assistants can ask questions concerning the animals to encourage students to think about adaptations.

When the rotation is complete, students pick up the animal skins and return them to the display shelf. Possible skins may include the following list.

Bat	Mouse
Beaver	Deer
Chipmunk	House
Ground hog	White-footed
Long-tailed weasel	Muskrat
Mink	Norway rat
Mole	Opossum
Eastern	Rabbit
Star-nosed	Raccoon

    Shrew
      Musket
      Short-tailed
    Skunk
    Squirrel
      Flying
      Fox
      Red
      Thirteen-lined ground
    White-tailed deer

B.   The Auction

Groups buy as many skins as they can and stay in the bidding as long as possible. Each group has a bidding power of thirty gold nuggets (thirty pieces of corn) and is given a small container with the thirty pieces of corn. Stress that a group needs to discuss an animal before bidding on it. The students have just examined are auctioned.

The auctioneer does not indicate the name of the animal for sale, but describes it by giving three

adaptations. After these are given, each group discusses ideas concerning which animal is indicated. A group, thinking it has the correct answer, begins bidding by having one of its members raise a hand. Once a group bids the highest number of gold nuggets, the auctioneer asks that group to name the animal. If the answer is correct, the group pays the amount bidden to buy the animal. If the answer is wrong, the group still must pay, but it does not get the animal skin. The bidding then starts again at a lower price.

Some animals may be sold as two for the price of one because some adaptations may describe more than one animal; therefore, more than one answer may be correct. The auctioneer should always mention eye placement and type of teeth since this helps students.

The following list offers animals with their adaptations which may be sold at an auction.

1. Rabbit or white-tailed deer

    a. Two for the price of one
    b. Flat teeth with incisors
    c. Eyes on side of head
    d. Tail held high as warning signal for danger

2. Chipmunk or thirteen-lined ground squirrel

    a. Two for the price of one
    b. Flat teeth with incisors

If you are a hawk, looking down for something to eat, you would have a hard time seeing this animal on the ground because it has a broken pattern of coloring on its back. If this animal is in a tree and you are a weasel looking up, this animal would blend in with the light-colored sky since its stomach is light-colored.

3. Weasel or mink

    a. Two for the price of one
    b. Eyes in front of head
    c. Sharp canine teeth
    d. Long, slender body to get in holes of prey

4. Muskrat or beaver

    a. Two for the price of one
    b. Eyes on side of head
    c. Flat teeth with incisors
    d. Flat tail used for rudder in water

5. Fox squirrel, red squirrel, flying squirrel

    a. Three for the price of one
    b. Eyes on side of head
    c. Flat teeth with incisors
    d. Tail used for blanket, balance, umbrella, and rudders in air

6. Opossum

   a. One for the price of one
   b. Eyes in front of head
   c. Sharp canine teeth
   d. Tail like a monkey's
   e. Thumb like a man's

7. Raccoon

   a. One for the price of one
   b. Eyes in front of head
   c. Sharp canine teeth

If you were an owl and a human walked below you, your first reaction may be to open your eyes to see what is making noise. Realizing the danger, you would close your eyes to a slit, since they stand out and give your position away. This animal has two adaptations to help hide its eyes: 1) its eyes are black, and 2) the background on its face is the same color as its eyes.

8. Short-tailed shrew, eastern mole, star-nosed mole

   a. Three for the price of one
   b. Sharp canine teeth
   c. Barely noticeably eyes on front of head

If you pull out a single strand of hair from your head, you find that the part by the roots is thicker than the other end. This animal has hair which is just the reverse; it is thin at the roots and thick at the other end. Because of this, the hair can be pushed both ways with little resistance. If a horse or dog is petted, the fur easily goes one way, but it is difficult to push in the opposite direction. The air of this animal can be pushed either way.

9. Skunk

   a. One for the price of one
   b. Eyes in front of head
   c. Sharp canine teeth
   d. Large warning signal on top

This ends the auction. Or if time allows, other animals with their adaptations may be sold. At this point, students should have a fairly clear understanding of the term ''adaptation'' and the importance of an animal's adaptations.

C. Matching Skulls and Skins

The instructor places all the skulls in a long line. Next, a study skin — not necessarily the skin which originally matched the skull — is placed behind each. Students then try to unscramble the skulls, placing them in front of the correct study skins.

This activity moves faster with more student involvement. If there are enough skulls and skins to have two sets, one at each end of the room with a group working on each, students can switch after a certain period of time and try their skills at the opposite set.

III. Conclusion

   A. What would happen to an animal that is not adapted to its environment?
   B. What would happen if a rabbit was born with eyes in front of its head or a fox was born with wide, flat teeth like a squirrel's?
   C. What special adaptations do we humans have that separate us from the rest of the animals?

Habitat Hunt

Objectives:

1. To help students understand concepts of *habitat* and *niche* and apply this understanding to local animal populations.
2. To help students recognize the five basic needs of animals that must be met by a habitat.
3. To help students associate various animal signs with the animals that make them.
4. To help students recognize how man's impact on the natural environment has affected habitats.

Equipment:

1. Basic track guides
2. Leader cue cards (See Figure 2.2.)
3. Clipboard, paper, pencils (optional)
4. Role-playing cards (optional)

       a. Mayor
       b. Factory owner
       c. Realty company president
       d. Unemployed worker
       e. Shop owner
       f. Animal (deer, rabbit, fox, etc.)
       g. Plants
       h. Elements (air, water, soil)

I. Introduction

   A. What are the five main things animals need to survive? (Food, water, shelter, air, and space)
   B. What does the word *habitat* mean?
   C. Tell everything you know about the place a squirrel lives.
   D. Why is this place so special to the squirrel?
   E. What are some important things the squirrel does in its habitat?
   F. The perfect place where a squirrel can find everything it needs is called its *niche*. How does a squirrel know this is its right place? What tells it this? (Instinct. It inherited information which helps it survive in its habitat as well as a body, from its parents.)

LEADER CUE CARD

I. Motivating Activities
  A. Home Inventory: Who can find the most shelters?
  B. Tracks Inventory: How many different kinds of animals live here? Which are the most numerous?
  C. Search for complete habitats. Include shelter, food, water source, and name of animal.
II. What is man's impact on the habitat?
  A. Impact good for the habitat. In what way?
  B. Impact bad for the habitat? In what way?
  C. No impact on the habitat.
III. Discovery Methods
  A. Follow Tracks: Animal may give a tour of its habitat.
  B. Role-play: Assign each student an animal to role-play. The student must find the best, most complete habitat for himself. The students then can share their homes with each other on a habitat tour.
  C. Sizing Habitats: Divide the group into pairs. Ask pairs to find complete habitats for animals.
     Pair 1 — Mouse-size and smaller
     Pair 2 — Chipmunk-size to rabbit size
     Pair 3 — Animals larger than rabbits
     Compare findings.

Figure 2.2

G. What would happen to a squirrel if it were dropped off in the desert or in the middle of a busy, big-city intersection?
H. We are going out in a few minutes to hunt for some real habitats. The animals often leave clues to help us find their habitats. What are some clues or animal signs we may find? (Tracks, holes, nests, broken twigs, food scraps, fur, skin, feathers, shells, droppings, paths, the animal, sounds of the animal, etc.)

II. Field Experience

A. Divide students into field groups. The size and number of groups depend on the class size and number of group leaders.
B. Field groups are sent to different areas, relatively near, to hunt for habitat. Time spent on the habitat hunt varies. Provide at least twenty-five minutes. Thirty to thirty-five minutes is ideal.
C. Leader cue cards, designed to assist the field group leader in conducting the habitat hunt, make the hunt exciting and challenging for students.
D. A secretary records all of the group's findings: types of animal signs, their natural communities, and their quantities. A rough map of the investi-

gated area is attached to the back of a clipboard and signs plotted on it with a grease pencil.

III. Conclusion

A. Draw a rough map on the board of the area investigated. Divide the map into communities. Ask each group to tell what they found in their areas. List and plot these findings on the map according to the community in which they were found. When all the groups are done, everyone has a good overview of what is in each animal's habitat. Use symbols to plot animals, tracks, homes, etc.
B. Environmental problems are discussed. This allows for application of habitat understanding to a current environmental issue, loss of habitat.

IV. Post-Experience Activities

A. Make plaster casts of animal tracks found around the school area. Turn these into art projects to be hung in the classroom.
B. Utilize cue card ideas in the habitats.
C. Do habitat field investigations around the school site and then create a habitat map of it, checking back on these habitats from time to time.
D. Provide more wildlife habitat areas at your school. Improve shelter by creating brush piles, plantings of low, bush shrubs. Save old logs. Set up bluebird, wood duck, and kestrel boxes for nesting. Set up feeding stations for winter birds and squirrels. (Do not abandon feeding stations once they are started.)

Ecosystems

Objectives:

To help students understand:

1. What is meant by *ecosystem*.
2. How fluctuations in populations directly result from a predator/prey relationship and food supply.
3. Ecosystems should be balanced, *balance of nature*.
4. The concept of interdependence of plants, animals, and non-living things in an ecosystem.
5. Energy, which powers the ecosystem, starts with the sun.
6. Energy travels from the sun to living things in the ecosystem through the food chain.
7. The concepts of *food web* and *food pyramid*.
8. What happens when one component of an ecosystem is destroyed.

I. Activities

The entire class consists of the following activities:

A. Food Chain Scramble

Concept: Food Chain

Students are put in groups of four or five. Each group is given a set of four or five cards with names or pictures of plants or animals. Each set must represent a complete food chain. Students line up in order of the food chain, having the animal or plant they would eat to their left. For example, cards containing plant juice, mosquito, frog, snake, and hawk would be held up by students who would line up in that order.

Equipment:

Approximately twenty six-inch by four-inch cards with pictures or names of various herbivores, carnivores and plants

Discussion:

What happens to the mosquito if we spray poison on plants? What happens to the frog that eats the mosquito? What happens to the snake and the hawk in turn? Even though we had no desire to hurt the hawk by spraying plants, what is the result of our destruction of plants? Why do animals get larger as the food chain moves up? Where does man fit into the food chain?

B. Eco-Energy Game

Concepts: Energy Flow, Food Chain, Food Pyramid

Explain that the path which all energy takes from the sun to living things is called a food chain. Break the class into four-member teams, giving each a set of food chain cards. Members then line up in order to make a food chain. Next, have the team members line up, one behind the other about twenty yards apart, with the teams in parallel lines. About twenty yards in front of each line is a full can of hickory nuts, acorns, etc., which represents the sun's energy source. Students also have cans by their feet which represent the energy used by plants or animals that cannot be passed on. They are called *waste cans.* The idea is to pass energy from the sun to the animal at the top of the food chain. However, before a student can pass energy to the next plant or animal in the food chain, he must drop a piece of energy into the waste can.

When the game starts, the first student runs to the supply bucket (the sun), picks out a piece of energy, runs back to his own waste can, and drops it in. He then runs back to the sun, picks up another piece of energy, and delivers it to the next student in the chain who drops it in the waste can. The first student immediately runs back to the sun for another piece, but again must drop it in his own can. He then runs to pick up another and delivers it to the second student, who delivers it to the third. The process continues until the last student in the chain, the carnivore, has acquired one piece for his can and a second piece to cross the finish line. The first team to have the last animal in their chain cross the finish line wins.

Equipment:

1. Energy pieces (hickory nuts, peanuts, poker chips, paper clips, etc.)
2. Small buckets
3. Food chain cards (set of four)
4. Line markers
5. Bases
6. Small tin cans

Discussion:

Who wasted the most energy in the game? (The plants.) Who is the most tired? (The plants.) Why are so many plants in our world? (To support all the animal life on earth.)

Have students again arrange themselves in order of their food chain and state some environmental problems which might eliminate parts of it. Such problems might be: 1) excessive grazing of cows which destroys all the grass, 2) insecticide which kill off all the grasshoppers from this field, 3) smog which blocks the sun from the area, or 4) all carnivores which are lost from excessive hunting or DDT poisoning.

C. Foxes, Rabbits, Clover Game

Concepts: Cycles of Population, Balance of Nature, Extinction

Have the class stand in a circle and count off by threes. Ones are foxes and take fifteen giant steps backward. Twos are clovers and take ten giant steps backward. Threes are rabbits and each puts one hand on the instructor who is in the middle. The object of this game is to *eat.*

Rules:

1. Rabbits must tag clovers before they are tagged by foxes or starve before time runs out.
2. Foxes must touch rabbits before rabbits touch clovers or starve before time runs out.
3. Clovers cannot move and must hold out their arms so rabbits can find them.
4. Each rabbit eats (tags) only one clover to be safe from foxes who can eat only one rabbit.
5. Only one rabbit can eat a certain clover, and only one fox can eat a certain rabbit.
6. The instructor says "go" then waits about thirty seconds to yell "stop." At that time everyone must freeze.

In the next game:

1. Clovers not eaten by rabbits remain.
2. Foxes and rabbits who tagged food survive.
3. Clovers, tagged and making new rabbit cells in rabbits' stomachs, become rabbits.
4. Rabbits, tagged by foxes, become foxes.
5. Rabbits or foxes who failed to tag their food starve to death and rot into the soil where clovers grow; thus, they become clovers.

After each playing time, students return to the circle, figure out what they are in the next game, and take their places. They should also observe the populations of foxes, clovers, and rabbits because fluctuations occur as the game progresses. Rabbits go up, but plants go down. When the rabbit population is higher, the fox population increases. However, several games later the foxes eat the rabbits and their population once again is small. Eventually, one of the three components of the game and the animals which depended on that component disappear. At the end of each game, have students predict what is going to happen during the next playing time.

Discussion:

What happened to the rabbit population when the clover diminished? What happened to the fox population when the rabbits diminished? What was needed to make this a more balanced ecosystem? If the foxes (or rabbits) became extinct, how could they be brought back? What are some plants and animals that are nearing extinction? What are some that are already extinct? Could rabbits or other animals ever become so numerous that they take over the world? How does nature keep things in balance?

D.   Pyramid Game

Concept: Food Pyramid

Three students, representing grass, kneel side by side. Two more students, representing grass-eaters such as rabbits, kneel on top. One student, representing a rabbit-eater such as an owl, climbs on top. Thus, a human food pyramid is formed.

Discussion:

Why does it take three people, representing grass, to support one owl? (The concept of energy loss is seen again.) Does an owl eat grass? What happens if the grass is pulled out? (The instructor reaches down and pulls on several students' arms on the bottom level.) If owls do not eat grass, why did the owl fall when the grass was pulled out?

E.   Power Plant

Concept: Element of Life

Explain that the box is a power plant for the entire ecosystem. Everything needed to keep the system running is contained in the box; it is the source of all the world's energy.

Equipment:

Box with soil and water inside

Discussion:

Discuss what might be in the box; then open the box, revealing the ingredients of soil, water, sun, and air.

F.   Food Web

Concept: Population Balance

Have students sit in a tight circle and distribute cards of animals to them. Their cards should be placed on the ground in front of them so everyone can see. Give one end of the yarn to a carnivore at the top of the food pyramid, who then selects something to eat from the circle. Run the yarn to that student, hook it around his index finger, and ask what he wants to eat. Continue to the bottom of the pyramid until the yarn is connected to the plants and sun. Take the continuous line of yarn back to a member at the top of the pyramid and start over, continuing the cycle until the yarn makes a crisscross pattern, connecting each student to at least one other.

Equipment:

1.   Set of twenty or more cards with pictures or names of various plants, herbivores, and carnivores
2.   Spool of yarn

Discussion:

What does this look like? Where does man fit into the web? Does he need these things? Do these things need him? Why is a web formed? Who needs whom? Who has the most? Who is needed by everyone?

The instructor assures all that he has no intention of hurting the animals, but wishes to chop down some trees for farmland. Then, he cuts the trees by slipping the string off their fingers. All students holding loose pieces of yarn are directed to let go.

What happened? Why? Why should people be careful when interfering with the balance of nature? What precautions can we take to prevent this kind of occurrence? What happens to an ecosystem if the sun is taken away? What happens if man is taken away?

G.   Snarf

Concept: Predator-Prey Relationships

Students are put in two even lines, twenty yards or so apart, facing each other, with boundaries indicated some distance behind each line. The leader designates one line predators, such as foxes, and one line prey, such as rabbits. Predators attempt to tag prey before the prey cross the boundary line behind them. If caught, prey join the predator team. Students never know who chases or runs until both lines are named. Some suggestions for naming lines include birds/insects, owls/mice, mosquitos/men, mosquitos/dragonflies, pike/minnows; minnows/insects, plants/grasshoppers, or plants/sunlight.

H. The Power

Concept: An Ecosystem

Students break into groups of two or three and receive a cardboard writing surface, pencil, and paper. They are then given *The Power* which enables them to create their own habitat areas. Keeping in mind what they discovered in the previous activity concerning food supply and balance of populations, they can include any natural objects or animals they choose. Students should not be encouraged to include fifty wolves with two deer, but should be encouraged to include insects if they have birds. Also, their areas should be designed so none of their animals starve. Groups work together to write their lists of ingredients or recipes. Each group can write a recipe for a different habitat, or all groups can work on the same. Students could draw pictures or maps of their areas.

Equipment:

Cardboard writing surface, pencils, paper

Discussion:

Students can discuss advantages and disadvantages of their ecosystems.

II. Conclusion

This class can be as simple or advanced as the instructor wishes. All or some of the games can be played depending upon the level of students.

III. Pre- or Post-experience Activities

A. Make a large food web display for the classroom.
B. Make food chains from the students' meal. Have them make a list of their bad environmental habits. Elect and monitor a bad habit of the week.
C. Break students into small groups to go into several habitat areas in search of animal signs, homes, and foods. Discuss the signs found in each area. What animals live in each? What animals live in more than one? What evidence of food chains was found? How many different kinds of homes were discovered? What animals live in one area, but feed in another?
D. Make a list of many different animals. How many ways can they be classified? Let students think of classifications and place them into categories. These could include predator/prey, land/air/water, animals with homes below/above ground, animals who build/steal homes, and animals who live underground in fields/forests/lakes/swamps.

E. Tell students they have just been given a universe and they have the power to create a world. They must make a master plan, listing the items for their world in the order they want them to appear. A length of time separates each item as it appears; therefore, each must be able to survive with only those things created previously. Thus, could man be created first? Could he stand alone while other things are being created? After making the master plan, students simply snap their fingers and their world begins to take shape. No corrections can be made once the process is in motion. Decide as a group if the world they created will progress or fade away and die.

Wildlife Management

Objectives:

1. To help students gain a basic idea of the processes involved in wildlife management and the contributions made to wildlife through sound management techniques.
2. To teach and help students understand the terms *limiting factor* and *carrying capacity* as they apply to wildlife management.
3. To familiarize students with the roles of hunting and trapping in wildlife management.

Equipment:

1. One plastic or paper cup for each student
2. Bag of popcorn

I. Introductory Discussion

A. What does every animal need in order to survive?

1. Air	4. Water
2. Space	5. Shelter
3. Food	

B. A piece of land usually has enough air and space for any number of rabbits, but consider the following:

1. If a certain area has enough food for two hundred rabbits, enough shelter for one hundred, and enough water for four thousand, how many rabbits could live in that area? (Only one hundred rabbits could survive since the rest would not have any shelter.)
2. If we create brush piles and briar patches to shelter four hundred rabbits, how many rabbits could live there? (Only two hundred could survive because the food would run out. Food is the factor limiting the number of rabbits in that area.
3. The factor limiting the number of animals that can survive in a given area is the *limiting factor* for that area.

II. First Game

A. This game shows how *limiting factors* work in the real world.

    1. Students stand in a large circle outdoors. Two students, selected to be parent rabbits, are given a paper cup to represent their stomachs. About three cups of popcorn are scattered on the ground inside the circle to represent vegetation, the rabbits' food. Mom and Dad rabbits are given thirty seconds to fill their stomachs. This should be easy since there is plenty of food. Afterward they put their popcorn back into the bag.

    2. Next, about seven cups of popcorn are thrown on the ground to represent grown vegetation, while four more students are chosen to be first-generation children. Along with Mom and Dad they must fill their stomachs in thirty seconds. This is a little more difficult, but there should still be enough food for all six rabbits.

    3. Another generation comes along, a little more food grows back (about eight cups of popcorn are thrown on the ground), and all the remaining students represent grandchildren. All the rabbits — Mom, Dad, children, and grandchildren — must fill their stomachs. This time only six or eight rabbits can fill their stomachs, while the rest starve as the ground is picked clean. Food is the limiting factor of the area.

B. Discussion

    1. Look at the area. Is there enough food to support even one rabbit? Probably all the popcorn was picked up by the starving rabbits. In the real world animals also eat everything possible before they eventually starve to death.

    2. How many rabbits were able to live before there was not enough food? The parents and four children had enough to eat; thus, land could support six rabbits. The number of animals the land can easily support is called the *carrying capacity* of the land. What is the present carrying capacity of this piece of land? (It is zero because it cannot support any rabbits until the food grows back.)

III. The Second Game

A. One student plays a fox. The rest of the students are rabbits.

    1. The first part of the game takes place in an open area where there are no hiding places (shelters). With eyes closed, the fox stands in one place and counts to ten. All the rabbits run, trying to hide. On the count of ten, all the rabbits must stop in place, and the fox opens his eyes. Rabbits which the fox can see by simply turning around are eaten. This includes nearly all the rabbits since there is no shelter.

    2. This time the game is played in an area with bushes and trees for hiding places (shelters). The same procedure is followed. This time the fox only eats four or five rabbits and the rest survive since they have shelter.

B. Discussion

    1. What was the *limiting factor?*
    2. How did it affect populations?

IV. The Wildlife Manager

A. The wildlife manager's job is to study limiting factors and carrying capacities. His goal is to enable the land to easily support as many animals as it can without upsetting other populations.

B. A wildlife manager has several options to make animal populations of a given area healthier.

    1. The carrying capacity can be increased by adding shelter, food, or water to the land area.

    2. Hunting or trapping by individuals to control the animal populations before starvation can be allowed.

    3. Natural predators might be introduced to the area to control the populations.

    4. Excess animals can be moved to a different area.

C. Draw a sketch of an area that consists of 250 acres of lake, 250 acres of forest, 250 acres of open field, and 250 acres of a ten-year-old field. (See Figure 2.3.)

    1. If we know that a rabbit covers only about four acres of land in its lifetime, where could a rabbit live in that area? (Only in the center where it can find food, water, and shelter. Even with 1,000 acres of land, only a few rabbits can live only in the center!)

    2. What could we do to allow more rabbits to live here? Have several students sketch some ponds, brush piles, open areas, canals, or whatever else the group might suggest. These kinds of changes are part of a wildlife manager's job.

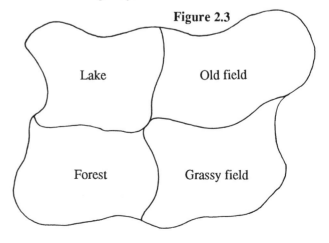

Figure 2.3

Lake

Old field

Forest

Grassy field

V. The Debate

Students are presented with an imaginary problem situation involving a piece of land that is overpopulated by a certain species of animal — deer, rabbit, muskrat, or other. This should be as realistic a situation as possible, perhaps one true-to-life situation concerning a nearby piece of land.

A. Have students list possible solutions: altering the land to increase carrying capacity, hunting or trapping, introducing predators, or removing animals.
B. Students gather in groups according to the decision they think is best. They then have a given length of time to organize their thoughts in support of their opinion and to select a group spokesperson.
C. Each group, in turn, presents its views and arguments in favor of its opinion. If any other group's opinion is challenged, that group is allowed a rebuttal.
D. All groups should have a chance to present their opinions.
E. Another way to set up a debate is to have all students sit in a large circle. In a smaller circle inside the big one, place four people: one in favor of hunting and trapping or some other such measure, one who is against it, one who is neutral, and the instructor. Only the people in the center of the circle can state their opinions. If anyone from the outside would like to become involved in the debate, they simply tap any of the four people on the shoulder and trade places. The new student then can state an opinion, question, or argument.

VI. Conclusion

A. The job of a wildlife manager often involves very difficult decisions.
B. These decisions often affect more than one species at a time.
C. Many decisions, especially those concerning hunting and trapping, are personal and are not necessarily always right or wrong.

In Cold Blood

Objectives:

1. To help students recognize the external differences between reptiles and amphibians.
2. To help students identify at least six different reptiles and three different amphibians.
3. To make students aware that reptiles and amphibians have adaptations for living in their environment, protecting themselves from enemies, and obtaining food.
4. To help students, through close contact and awareness of these animals, gain a better appreciation of reptiles and amphibians, which are often held in low esteem.

Class Materials:

1. Aquariums which have appropriate space, food, and water
2. Snapping turtle
3. Box turtle
4. Three painted turtles, map turtles, or pond sliders
5. Three leopard frogs or any other type of frog
6. Three American toads or fowler toads
7. Three garter snakes or any other non-poisonous snake

Optional:

1. Musk turtle
2. Soft-shell turtle
3. Spotted salamander

Class Arrangement:

When setting up class, it is important that each aquarium have a label stating the kind of animal and whether it is a reptile or an amphibian. Caution signs should be on the aquariums containing animals which may bite. Snapping turtles, musk turtles, soft-shell turtles, and snakes fall into this category.

Aquariums should be placed on the floor of the classroom in a large enough circle so that students can sit in a circle inside the circle of aquariums. All animals which may bite should be placed together, and frogs and salamanders should be placed side by side. The rest of the animals can be placed in any order around the circle.

I. Introduction and Rotation

A. Name different types of reptiles found in this area. (Snakes, turtles, lizards)
B. Name different types of amphibians found in this area. (Frogs, toads, salamanders)
C. Rotation — Have each student pick a partner and go to one of the aquariums. Tell them they have about thirty seconds to examine its contents before rotating in a clockwise manner to the next aquarium. Encourage students to examine the label on each aquarium, find the animal's name, and check to see if it is an amphibian or reptile. If desired, they may handle the animals.
D. Handling — Some animals must be handled with special care. Demonstrate this to students before they begin the rotation.

    1. Snapping turtle — Can be held only by the tail.
    2. Soft-shell turtle — May be picked up and held by the back of the shell.
    3. Snakes — May be handled gently, with both hands. Snakes in captivity are used to being handled and are quite docile.
    4. Frogs — May be handled, but only with wet hands. They can be dipped in a bucket of water which is placed between the frogs and salamanders.
    5. Salamanders — May be handled, but only with wet hands.
    6. Box turtle — Very safe to handle.

7. Painted turtle — Very safe to handle.
8. American toad — Very safe to handle.

Students should be urged to handle the animals with tender, loving care. Rotation should last about five or six minutes. Students never leave the circle and should end where they started, ready to begin the second part of class.

E.   What is the external difference between reptiles and amphibians? (Reptiles have scales and amphibians have smooth skin.)

Part I helps familiarize students on their own level with the characters in class. Students can touch, if they want; but they may choose otherwise. The second part of class requires that blindfolded students touch four different animals. For some this is more than they expected; therefore, the instructor's approach of this is very important to the success of the class.

II. Adaptations

Remaining in the circle, students are blindfolded and have their hands behind their backs. The instructor, with the help of leaders or teachers, goes behind the students and places one type of animal in their hands. Without talking, students should feel the animal and determine its type. If there are three toads, three snakes, and three painted turtles, the other leaders can take them around when the instructor does to save time. Once students guess the animal, the instructor discusses its adaptations. A questioning strategy is used which encourages students to determine various adaptations and their uses. This part of class should not be a lecture.

Each animal is placed in the middle of the circle for discussion.

A.   American toad

1. What happens to a dog when it eats a toad? (It foams at the mouth and gets sick.)
2. Why? (The toad is poisonous.)
3. Where is the poison on the toad? (In the warts.)
4. Why should you be sure to wash your hands after handling the toad? (You might touch your eyes or mouth, getting poison in them.)
5. How do toads protect themselves from their enemies? (With poison.)
6. Snakes are not affected by the poison, so how could a toad get away from a snake? (Hop away.)
7. Do toads hop as well as frogs? (No.)
8. Have a student hold a toad by the two back legs so that the rest of the class can see what happens. What is the toad doing? (Puffing up.)
9. Where would a snake grab a toad it wanted to eat? (By the legs.)
10. What can the toad do so the snake cannot eat it? (Swallow air and puff up so it cannot fit into the snake's mouth.)

11. Have any of you picked up a toad outdoors? What does it often do? (Wets.)
12. What do you do when that happens? (Drop it.)
13. What would a toad's enemy probably do? (Drop it.)
14. Imagine trying to eat a toad. First, it tastes terrible. Next, it puffs up to get caught in your throat, and then it wets all over your face. Do you want to eat a toad?
15. Do toads shed their skins? (Students usually do not know. Toads do shed; however, they eat the skin as they shed. For this reason toad skins cannot be found lying about like snake skins.)
16. How can you tell a female toad from a male toad? (Students usually do not know, but have a good time guessing. Toads cannot even tell the difference. When they mate, a male toad gets on the back of another toad. If it lets out a chirping noise, it is a male. Demonstrate by gently squeezing the sides of a male toad. Also, let a few students try it.)

A list of other animals and facts about them follows. Be sure to develop a questioning technique which lets students bring out these facts.

B.   Garter snake

1. A snake sticks out its tongue to gather odors which it tastes as the sense is computerized in the brain.
2. A snake can eat fairly large animals because its jaws unhinge and open much wider than other animals' jaws which are locked.
3. A snake's teeth slant toward the back of the mouth so animals can slide in, but cannot wriggle out because they are fishhooked on the teeth. Thus, if you are bitten by a snake, do not pull away.
4. A snake gets cloudy eyes when it is about to shed because the skin over them sheds as well. Cloudy eyes are caused by the skin loosening over them.
5. A snake sheds because it outgrows its skin, which does not grow with it. The skin simply splits and falls off after the snake grows a new layer underneath.
6. A snake is cold-blooded; therefore, its body temperature varies with the outside air temperature.

C.   Painted Turtle

1. Lay the turtle on its back to show students it flips over by itself.
2. Land turtles, such as the box turtle, have more box-like shells than water turtles. For instance, the painted turtle has a more streamlined shell. Water turtles have webbed feet to aid them in swimming.
3. These turtles protect themselves by withdrawing into their shells.

4. Snappers protect themselves with powerful jaws and a long neck which can be extended very far. Students can see this if the snapper is held by the tail. Snappers also flip over very fast from an upside-down position.
5. Musk turtles and soft-shell turtles bite to protect themselves since their shells are soft and they cannot withdraw into them.

D. Leopard frog

1. Place the frog in the center of the circle. Note how far it jumps. This is necessary for it to escape.
2. To protect its mucous membrane, wet your hands before handling a frog. If this membrane is broken, germs and infection can set in.
3. Frogs have an extra eyelid for protection under water. If gently touched with a finger just above the eye, it will lower.

III. Conclusion

Save five to eight minutes at the end of class so students can go back and handle their favorite animals. Again, stress proper handling.

IV. Further Ideas

A. Make sketches from personal observations, noting patterns and colors.
B. Discuss poisonous snakes found in North America. Should they be killed?
C. Maintain reptiles or amphibians in the classroom. Record the food accepted, methods employed in eating, changes in coloration, shedding of skin, and general habits.
D. Take photographs of reptiles or amphibians.
E. Discuss superstitions and common fears that people have about reptiles and amphibians.

Squirrelympics

Objectives:

1. To help students comprehend through first-hand experience the various adaptations of squirrels.
2. To help students understand the habits and nature of squirrels.

Equipment:

1. Wooded area with large trees and grassy ground cover, with at least five or six trees having a six-foot or more circumference
2. Bag of 250 to 300 peanuts
3. Twenty-five to thirty cellophane bags, six-inch size
4. Squirrel study skin or picture of a squirrel
5. Large tree about seven feet around
6. Three logs lying parallel on the ground about three feet apart

7. Log or beam about four inches in diameter and about twenty-five feet long
8. Roll of masking tape

I. Introduction

Discuss the types of squirrels inhabiting the area, such as the fox squirrel and the red squirrel which are two common types. Compare the characteristics of these two and include characteristics of any other squirrels students know. Mention the following adaptations:

A. Teeth — What do squirrels eat? (They are herbivores; therefore, they eat mostly nuts, pine cones, seeds, and plants.)
B. Coloration — Note that squirrels have a darker color on top and a lighter color underneath. Why? (Predators looking up see the lighter side which blends with the sky, while those looking down see the dark side which blends with the trees and ground.)
C. Paws — Squirrels have four claws and no thumbs. Why? (Thumbs may not be needed with claws. Perhaps, they have more dexterity and speed without them.)
D. Eyes — As herbivores and potential prey, squirrels need big eyes and good peripheral vision to watch for foxes, hawks, and other predators which might be lurking about. Planteaters' eyes are located on the side of the head so they can see around them. Predators eyes are located in front so they can see their prey and have binocular vision which gives them a better judgment regarding space and distance.
E. Tail — What is the purpose of a squirrel's tail? (It helps to keep the squirrel balanced when scampering along branches, and it protects it from wind and weather.

II. Eat or Be Eaten Game

A. Squirrels

1. Students become squirrels. Tape together the thumbs and forefingers of both hands, simulating squirrels' paws, which have no thumbs.
2. Squirrels seek to survive the winter. They must gather peanuts from the ground and hide them for their winter supply. Give each participant a cellophane bag to hide their peanuts.
3. About three hundred peanuts are scattered about the area, which includes brush and several large trees.

B. Rules

1. Squirrels carry only one nut at a time; thus, students can take only one at a time.
2. Squirrels' nut caches are not immune to nut swiping by fellow squirrels; therefore, students may take nuts from other students' bags. They can steal only one nut at a time and may not

take a nut if the owner is within two paces of the bag.

3. Students hide their bags anywhere in the designated area, return to the starting point, and wait until the instructor commands them to start. Students then must gather as many nuts as possible within a five-minute time limit.

C. Results

1. When commanded to stop, squirrels, with their bags, return to the starting point.
2. Any squirrels having less than five peanuts starved during the winter.
3. Compare other squirrels' quantities.
4. Discuss strategies.

D. Second Year

1. Introduce a new element, pursuers, to the game: hunters, predators, and disease. Have leaders or teachers play these roles. Fifteen seconds after squirrels start gathering nuts, leaders or teachers, walking only, move through the nut-gathering area, touching squirrels within their reach. Squirrels which are tagged are injured and must sit in the recovery room for one minute. If squirrels are tagged twice, they are dead. Squirrels can escape their pursuers by wrapping their arms around a tree that has a diameter larger than their arm span.
2. After this, squirrels hide their bags a second time and the instructor again scatters the peanuts.
3. After another five-minute year, call the squirrels back. Discuss strategies, frustrations, and number of nuts collected. Squirrels must have five nuts to survive.
4. Discuss squirrel senses and adaptations which were useful such as peripheral vision, the ability to climb trees, agility, etc. Emphasizing the adaptations of squirrels is a crucial aspect of the class.

E. If time permits, there can be a third year of nut gathering because students are better squirrels and enjoy their improvements. This time include a hunter who can shoot predators and a disease which can kill the hunter.

III. Squirrelympics

This series of challenges shows how the skills and adaptations of squirrels compare to our skills. Explain each game, then allow students to go to stations. A leader should be at each station to explain and supervise, while students rotate through the desired stations.

A. Nut Gone — Students each get one peanut. Have students hold their peanuts out in front of them; and on command, have them race to see who can shell and eat a peanut first. Remember, the thumbs are still taped!

B. Hanging On — Choose a tree about seven feet around. Students must jump onto the tree and hang on any way they can. The student who hangs the longest wins.

C. Tightrope — Anchor a four-inch wide beam or log to the ground so it cannot roll. Students must crawl along the length of the log without falling. A log raised several inches adds to the drama, but spotters may be necessary to insure safety.

D. Tree-to-Tree — Stabilize three logs which are lying on the ground parallel to one another and are about three feet apart. Logs should be approximately eight inches in diameter and three feet in length; also, they can be staked in place or partially buried to prevent them from rolling. Have students, staying balanced on each, jump from one log to the other without touching the ground.

IV. Conclusion

Discuss how it felt to be squirrels. Discuss why humans may fail if they must live like squirrels. What would happen if squirrels had to live like rabbits or moles? How do squirrels' specific adaptations help them?

Horse Sense

Objectives:

1. To acquaint students to the fact that horses have changed over the years.
2. To help students understand how and why adaptations have developed in the horse.
3. To teach students about safety, the proper way to handle and move around horses.
4. To prepare students for the trail ride by teaching them the basics of control in riding, mounting, walking off, turning, and stopping.

I. Historical Background

The earliest horse existed during the Lower Eocene Period, some fifty-five million years before man. The eohippus (dawn horse), also called hyracotherium, was a little more than twelve inches tall, had four toes in front and three in back, and was a browser with a short neck and small head. Upon its discovery in the nineteenth century, it was so different in appearance from its descendents that scientists had great difficulty identifying it as the predecessor of the modern horse.

The four toes supported the eohippus on the marshy land where it lived. As conditions changed, the center toes evolved into one large hoof, while the outer toes shrank into vestigial organs and no longer reached the ground. As the climate and terrain changed and grassland appeared, the neck of the eohippus lengthened and grinding teeth developed, enabling it to become a grazer.

The eohippus originated in North America and migrated north into Asia and Europe across the land bridge of the Bering Straits, disappearing from North America. The eohippus died out forty million years ago because it did not adapt to changing geological conditions. It was succeeded by the orohippus and then the ephihippus, which had a very similar skeletal structure but increasingly more efficient teeth. Next came the mesohippus, a three-toed browser which evolved into the merychippus, a three-toed grazer with longer grinding teeth. By the Lower Pliocene Period the phiohippus, the first one-toed grazer which was fully hoofed and three times the size of the eohippus, had developed. By the time homo sapiens arrived, the phiohippus had developed into the equus which stood thirteen hands (52 inches) high.

The equus, which originated in North America, migrated to South America and north to Asia, Europe, and Africa. It became extinct in the Americas about eight thousand years ago. However, numerous species, exclusive ancestors of the modern horse, developed in Europe, Asia, and Africa due to various climates and terrains. None of the ancients were actually horses by today's definition which states that horses must be over 13.2 hands high. Those measuring less are considered ponies.

In modern times man began to breed horses to obtain various qualities such as size, strength, endurance, hardiness, beauty, athletic ability, and disposition. Now a great many breeds are on record, some almost indistinguishable. The largest, the shire, was the medieval night's horse which stands about nineteen hands high. The smallest, but the strongest for its size because it can pull twice its own weight, is the shetland, which stands about ten hands high.

II. Discussion

The first part of class covers some of the above historical background. A discussion concerning how horses changed may help students understand such concepts as survival of the fittest, natural selection, and evolution.

It is important that students understand that these changes did not occur within generations, but over million-year periods. For example, the eohippus did not suddenly lose a toe; four toes were an asset, supporting the early horse in the swampy land where it lived. Perhaps horses, which fell prey to meat-eaters if they did not run, ran faster as the land changed by rolling forward onto the tips of their toes. Thus, those which were remained survived because of *natural selection*, the process by which nature selects animals that best fit their environments, is appropriate.

As horses walked more on the ends of their toes, their back toes barely touched the ground. Succeeding generations had no use for the outside toe, which shrank into the side of the leg over millions of years. The other three toes fused to form a hoof. Horses which could not adapt to the changing conditions died off. Thus, the term *survival of the fittest* describes animals which best fit into their environment and have the best chance at survival —animals which are able to pass survival

characteristics onto offspring. This process accounts for the adaptations of all animals.

Horse sizes changed since larger horses always had the advantage. They were able to survive and give birth to some larger offspring and, possibly, some smaller offspring which were usually killed. It is helpful to ask students if any of them have brothers or sisters who are taller than their mothers or fathers. What kind of children are they likely to have? Just as a five-foot parent would not have a child eight feet tall, neither would an eohippus have grown from twelve inches to sixty inches in a single generation. Changes in adaptations of any importance took millions of years.

This introduction is mostly discussion; therefore, it should not take more than fifteen to twenty minutes. Following the discussion students proceed to look at a horse and name various adaptations that have helped it survive over the years. As they name each adaptation, the instructor challenges students to figure out why an adaptation is beneficial.

It is important to stress that the major part of the horse's development took place *before* man appeared. The horse, for example, did not develop a strong back to carry man, but man rides the horse because it has a strong back. The horse's development came as a response to its environment and predators. For example, a horse would kick or bite in response to instinct, which told it enemies were stalking it. It also is interesting to point out that man has interfered with the horse's evolution, since random breeding now is rare; therefore, further changes in the horse are not a result of natural selection.

The following are some adaptations which students may notice and explain:

A. Ears — Because horse's ears are tall and move easily from side to side and back and forth, horses have very good hearing which enables them to hear other horses over a mile away. They become quite nervous on windy days since their hearing, upon which they depend heavily, is impaired.

B. Eyes — Large, wide eyes, which set on the sides of the head like those of other planteaters, allow horses to see nearly all around them, but not directly in front or behind. Therefore, we should always approach horses from their sides, within range of vision. Students can see the way horses see by placing their hands flat between their eyes and looking around, closing each eye alternately. Horses see one picture out of one eye and another out of the other eye. The two pictures do not blend together. Meat-eaters' eyes are located on the front of their heads; thus, the two pictures blend together, resulting in binocular vision. Herbivores need eyes on the side so they can watch for meat-eaters that sneak up on them; but as a result, they cannot see depth or judge distances well.

C. Nostrils — Horses' nostrils are large and very sensitive. A horse can pick up odors from over a mile away, but cannot distinguish as many different odors as humans.

D. Teeth — Horses have typical planteater teeth, developed for biting off and chewing grass in the back of the mouth. Wide, flat teeth are in front, a space with no teeth (where bits rest when horses are bridled) follows, and grinding molars are in the back. Horses' teeth never stop growing, but continually wear down; therefore, horses' ages can be judged by their teeth: the more triangular the top surface of the front teeth, the older the horse. Also, the older the horse, the more front teeth protrude in a bucked-toothed appearance. Another interesting fact is that only male horses have an extra tooth located right behind the front incisors on the top and bottom. It is helpful to have several skulls of various planteaters and meateaters so that teeth and eye placement can be compared.

E. Muzzle — This very sensitive part has long whiskers to help horses sort out objects. It also gives horses a good sense of taste and the ability to easily identify foreign material. Horses have very tactile lips which can reach out, grasp, and hold objects in a way similar to that of an elephant's trunk.

F. Neck — The long and muscular neck allows horses to comfortably graze high into the trees. It also gives great force and strength while striking enemies with their teeth.

G. Size and Weight — Both are advantageous for protection, speed, and endurance.

H. Skin — Horses' skin is so sensitive that it can flick off flies which land to lay eggs.

I. Tail — Developed to ward off flies and other insects, it also acts as a shield to protect the vital areas beneath.

J. Mane — Helping to protect vital nerves and blood vessels which run along the topside of the neck, the mane keeps that area warmer in winter and shades the neck in summer.

K. Legs — These are powerful, yet very fragile. A broken leg can render a horse useless simply because an injured leg cannot bear the 1,000-pound weight of the horse's body. Legs are efficient in giving the horse a means of protection and great speed.

L. Hoof — The hard hoof protects fragile inner bones and cartilage and can do great injury to predators or other horses. The hoof also enables the horse to dig beneath snow and ice to the dry grass below. It is made of the same material as our fingernails; similarly, it must be trimmed periodically to prevent over-growth and chipping. The outer material is dead and senseless; the inside, like the quick of our fingernails, is live material.

M. Frog — Being the V-shaped cartilage on the bottom of the foot, the frog works like a miniature heart to pump blood back up the leg. It keeps the foot warm because it circulates the blood and acts as a shock-absorber. The outer layer of the frog is dead material.

III. Riding Horses

A. The instructor shows students how to approach, lead, mount, walk off, steer, stop, and dismount horses.

B. Next, a description of the trail ride is given so students have an idea of what to expect.

C. At this point, it is wise to let students who have never ridden to mount, walk horses around a bit, and dismount so they are more comfortable when they participate in the trail ride. Other students can do the same if there is time.

IV. A Few Basic Rules to Cover During Class

A. Slowly approach a horse from the side. Never run in the stable area.

B. As you approach a horse, speak and touch it so it knows where you are at all times.

C. Walk close, with your hand sliding across its rump, when crossing behind a horse.

D. Watch out for your own toes; a horse does not.

E. Do not enter a stall, unhook, mount, dismount, or do anything with a horse unless instructed to do so.

V. Posters as Visual Aids and Items of Interest

These are optional and can be added to the barn or classroom. Posters might include:

A. Sketches of horses' facial markings: blaze, star, stripe, snip, baldface

B. Horse drawing with the main parts labeled: withers, poll, hock, fetlock, dock, barrel, cannon, hoof, etc.

C. Drawing of the hoof with parts labeled: wall, sole, frog, white line, quarters, toe, etc.

D. Drawing of a saddle and bridle with parts labeled.

E. Drawing of various horses in the evolutionary states: eohippus, protohippus, mesohippus, modern equus.

F. A list and description of colors: sorrel, palomino, bay, dapple, dun, buckskin, pinto, etc.

G. Vocabulary list relating to horses with the definitions: stallion, gelding, mare, filly, foal, colt, yearling, horse, pony, etc.

H. A list of all the barn rules and trail ride rules

I. A list of the types and descriptions of the various breeds

Horse Ride

Objectives:

1. To have students participate in a unique outdoor activity.
2. To give students, under careful supervision, a chance to try something new and exciting and, possibly, increase their feelings of independence and willingness to try new things.

I. Introduction

A. Review the material covered in the Horse Sense class regarding basic control of horses: moving off, turning, and stopping.

B. Cover the details of the trail ride procedures.

  1. Keep horses in line and try not to let one horse pass another.
  2. Try to keep some spacing between the horses.
  3. Only walk the horses, unless otherwise directed.

C. Consider a few other safety issues.

  1. Students should not use stirrups unless they have proper riding boots; tennis shoes and hiking boots can cause a dangerous situation.
  2. Hard hats are recommended for all students on all trail rides.

II. The Trail Ride

A. Students are taken in class groups which may number up to twenty students. This is done safely as long as at least three staff members who are competent in dealing with horses accompany the ride.

B. The actual ride is only about half an hour. This is after allowing time for mounting, beginning the ride, and dismounting them at the end.

C. A walk trail ride is usually best due to the students' lack of experience and the number going on the ride.

D. This option allows half the students to take a ring lesson while the other half has a ground lesson of some sort. The two groups then trade positions. The advantage is that fewer horses are needed and fewer mounted students increase safety.

III. Conclusion

A. At the end of the ride, unattended students cannot dismount, hook up, or move around the horses.

B. Many students get excited about returning to the barn later to help clean stalls and brush a few horses.

Bibliography

Alexander, Fichter. *Ecology, A Golden Guide.* Racine, WI: Western Publishing Company, 1973.

Babcock, Harold L. *Turtules of North Eastern U.S.A.* NJ: Dover Publications, Inc.

Consant, Roger. *A Field Guide to Reptiles and Amphibians.* Boston, MA: Houghton, Miffling, 1978.

Dickerson, Mary C. *The Frog Book.*

Klots, Elsie B. *The New Field Book of Freshwater Life.* NY: G. P. Putnam's Sons, 1966.

Needham and Needham. *A Guide to the Study of Fresh-Water Biology.* San Francisco, CA: Holden Day, Inc., 1962.

Reid, Zim and Fichter. *Pond Life: A Guide to Common Plants and Animals of North American Ponds and Lakes.* NY: Golden Press, 1967.

Smith, Dr. Hobart M. *Snakes as Pets..* T.F.H. Publ., Inc.

Van Matre, Steve. *Sunship Earth.* Martinsville, IN: American Camping Association, 1980.

Outdoor Feelings

chapter 3

Discovery Hike

Objectives:

1. To bring into focus environmental concepts such as communities, seasonal changes, succession, diversity, etc., through first-hand contact.
2. To incorporate a number of different topics in a single class to gather ideas from other classes.
3. To give students a sense of accomplishment by having them complete a rather long hike.
4. To encourage social interaction.

I. Pre-experience

Before coming to camp, some simple hike activities or scavenger hunts can be conducted on the school site. These might include:

A. *Shape Hike* — Students look for objects with particular shapes (round, square, linear, hand-shaped, heart-shaped, etc.).
B. *Color Hike* — Students record or collect objects with certain colors.
C. *Alphabet Hike* — Students find an object for each letter of the alphabet.
D. *Sound Hike* — Students are given a list of sounds to find (birds, children laughing, trees rustling, etc.).
E. *Scavenger Hunt* — Students are given a list of objects to record or collect (a seed, an animal bone, piece of litter, a track, something an Indian might use, etc.).

II. Preparation

A. The length of the hike may depend upon weather conditions; but for class effectiveness, it should be offered when the weather allows a two-to-three-mile, two-hour, hike.
B. Everyone should be properly dressed, wearing comfortable shoes or boots. Rain gear and an extra coat or sweater should be taken if there is a possibility of weather change.
C. Cameras, binoculars, hand lenses, field guides to flowers and birds, and insect repellents may be taken. The hike is designed as a time of discovery rather than a race to the finish line; therefore, casual walking is encouraged.

III. On the Hike

A. *Meadows* — Travel through meadows and notice the many types of weeds and flowers. How do they spread their seeds? A variety of insects in the grass, such as spiders, grasshoppers, katydids, and others, can be looked at closely. Goldenrod galls can be opened to reveal small grubs, which bore out in the spring and become flies. Many edible plants are in the meadows. For safety and ideas on edible plants, see the Incredible Edibles class.
B. *Forests* — This is a good spot for birds and squirrels; so keep an eye open for wildlife. Point out nests and signs of birds in trees. Take a closer look at some spring wildflowers. Why do they bloom before the trees leaf out? In the fall, watch for a wide variety of mushrooms. Try to picture how the land looked in 1900.
C. *Marshes* — Marshes around a lake abound with wildlife. Keep an eye open for turtles, frogs, and snakes. Encourage students to observe and gently handle the animals, making sure they return them to their homes. Marshes are also excellent sports to observe shore birds and waterfowl; so many types of ducks and geese can be seen. Watch for these and hawks which may appear overhead. Many different types of wildflowers, including marshmarigolds and the rare fringed gentians, are in the marshes.
D. *Lakes* — A lake's history, how it was formed, where the old shoreline may be, and what now is happening to the lake may be discussed when walking along it. For more information on this topic, see Succession Study class.
E. *Pastures* — The hike may go past livestock pastures. Remember to respect animals by moving slowly and remaining quiet. Pastures are also good spots for mushrooms, so you may want to look for large (ten inches or so) puffballs. Also watch for birds, crows, pheasants, hawks, groundhogs, and other animals. If desired, compare pastures to ungrazed meadows, discussing the impact of livestock on the land.
F. *Roads* — Your hike may include a section of road; therefore, before stepping onto it, discuss safety. Remember to walk on the left side, facing traffic, so approaching cars can be seen. Walk on the side of the road rather than on the pavement. The

roadside is a good place to find wildflowers, edible plants, and dead animals. Discuss the impact a road might have on animals and their environment.

IV. General Hike Activities

These may be used at any time during the hike to keep the group's interest and spirits high between stops at points of interest.

A. *Songs* — Singing is lots of fun and its rhythm makes hiking easier.
B. *Sensory Bingo* — Cards can be made for a bingo game, which is played while hiking. Students must notice things like a flower blooming, birds singing, a feather, a squirrel's nest, etc., crossing these items off their cards until each has a straight line drawn.
C. *Pony Express* — Divide the group in half, forming two long parallel lines with individuals spaced far apart in each line. Start a message with the first person in each line, who then runs and whispers the message to the second person in line. The second person runs to the third, passing the message, and so on down the two lines. The first group to get the correct message to the other end of their line wins. Explain that Indians passed messages from memory over long distances.
D. *Points of Interest* — Divide the group in half, having everyone pick up a similar object (rock, stick). The first group goes ahead, placing their objects in the trail at points of unusual interest, while the other waits. The second group then follows the same trail, stopping at each point of interest to see if they can spot the interesting feature. After the two groups rejoin and compare notes, the roles are reversed.
E. *Snake Walk* — First have the group form a single line then explain that when you pass your hand over them, they become part of a giant snake. Once part of the snake, they must move with the rest of the body, following persons in front of them. They no longer can talk, but only hiss like a snake. Walk down the trail in a zig-zag snake pattern, playing a type of follow the leader.
F. *Friends* — As you hike, have students pick up a friend (any non-living object which catches their attention). At the next stop have them introduce the friend to someone else. The friend then must be replaced along the trail with the comforting thought that the next time students come through the area, a friend is waiting.
G. *Trust Walk* — Have the group form pairs. During the next part of the trail have sighted partners lead others whose eyes should be blindfolded or closed. Sighted partners may wish to share objects of unusual textures or smells with their blind partners. After reversing roles during another part of the trail, discuss what it was like to be blind.
H. *Cookouts* — A group can have a cook-out lunch at a designated spot, halfway through the hike, and finish the hike in the afternoon.

Enchanted Forest

Enchanted Forest, having a variety of activities, allows students to experience the out-of-doors by extensively using the five senses. Students, developing a sense of identity with their environment through personal involvement, are encouraged to be alert and aware of what is happening in the natural world.

Objectives:

1. To help students become more sensitive to their environment.
2. To increase students' powers of observation.
3. To increase students' ability to relate to their individual lives.
4. To stimulate students' imagination.
5. To help students feel more free to communicate their feelings.
6. To sharpen students' senses.

I. Blindfolded Hike

A. Equipment

1. Twenty-five-meter rope
2. Blindfolds, torn sheets or rags
3. Clipboards or pieces of cardboard boxes to provide adequate writing surfaces
4. Pencils or crayons
5. Paper

B. Procedure

1. Before beginning the actual hike, lead a discussion about the senses: ''What are the senses? Which senses do you use the most? How does a blind man compensate for his loss of sight?'' If desired, mention the dependency upon eyes and how the other senses often are overlooked.
2. Create the mood, a feeling of excitement: ''We are about to enter a special place where your senses will be stretched. First, we are going to eliminate one of your senses so you can increase the perception of your other senses.''
3. Spread the rope out and place students along it about a meter apart so they do not trip over each other. Tell the students not to move up or back from their places along the rope and to let the rope pull them. Also assure them that they will not be led into a tree.
4. Silence. Tell them this is a quiet hike and they should not talk, but listen to the sounds of the forest, smell the odors, and feel the ground beneath them as they walk along.
5. As students perceive the different senses, encourage them to let their minds take them wherever they want to go through the *Enchanted Forest*.
6. Occasionally instruct them to stop and listen, focusing on sounds near and far. Let them

reach down and touch the ground, each picking up a handful of forest litter. Have them smell, feel, and crunch the litter in their hands, sensing the sound. Then proceed as before.

7. Along the way, pass items down the line for students to touch, smell, eat, or hear. Suggestions: sumac berries, dandelions, burrs, acorns, pine needles, insect galls, snow, mud, sand, fur. Have students point to the sun, a sound, and their starting place.

8. After walking to a destination far from other groups, have students stop, face the direction of your voice, and sit down. With blindfolds still on, have them dig a hole, using all of their senses to experience the soil: "Feel it, smell it, rub it on your faces." Have students guess where they are. "What clues do you have?" (While talking, pass out paper, pencils, and clipboards.)

9. Let students take off their blindfolds. Have them write about:
   a. their different sensory experiences and feelings while being blindfolded; or,
   b. where their imaginations led them through the *Enchanted Forest*.

10. Let students share what they wrote.

II. What's My Leaf?

A. Separate children into small groups of five or six and have them sit in small circles as groups.
B. Have them close their eyes.
C. Give each group leaves from five or six plants so each student can make a friend.
D. Have students use all of their senses except sight to learn as much as possible about the leaves.

  1. Touch with tips of fingers.
  2. Explore edges.
  3. Note texture, shape.
  4. Touch on back of hands.
  5. Smell it.
  6. Taste with tips of tongues.

E. After one or two minutes, have them place their leaves in the center of the circle.
F. Have students open their eyes and find their "leaf friends."
G. Have each explain how the leaf was chosen from the pile. Have students relate their sensory findings and catalog the variety of characteristics.

III. Trust Walk

Have students pair off and take a trust walk in which one person's eyes are closed while the other leads the blind person around in silence. Afterward, have them share their feelings with their partners. Students then change positions.

IV. Conclusion

Sit in a circle and discuss the group's experience in the *Enchanted Forest*. Discuss how it feels to be blind and things that students noticed about their senses through the class.

Creative Expression

Objectives:

Through a series of enjoyable and interesting activities, students are encouraged:

1. To take notice of the natural world around them.
2. To use numerous, creative ways to experience, interpret, and express the natural world.

I. Drawing Sounds

A. Students need paper, crayons or chalk.
B. Have students close their eyes and listen to the natural sounds around them. Talk about how sound is an invisible phenomena. Then ask what the sound would look like if it could be seen (color, shape, size, etc.).
C. Have students draw a picture of the sound with a crayon.
D. Other senses can be substituted. Emotions and feelings also may be used.
E. This activity also can be done in the classroom as a pre- or post-outdoor education experience.

II. Odors

List at least three smells that you can identify in your area of land. Describe the smell in two or more words.

III. Essence Hunt

A. Go out and find a natural object that reminds you of:

  1. peace
  2. your group
  3. yourself
  4. your friend

B. Share reasons for your choice. Can you think of any other ideas for this activity?

IV. Poetry

A. Make a cinquain poem about any natural object in the area. A cinquain poem is any poem which has the following constructions:

  1. It is five lines in length.
  2. The first line is the title and consists of only one word.
  3. The second line is a description of the title and contains two words.

4. The third line tells of some action and contains three words.
5. The fourth line expresses an emotion and contains four words.
6. The fifth line is made up of another word picture for the title and is one word.

B. Make a haiku poem of a natural object. A haiku follows this construction:

1. The first line is made up of seven syllables.
2. The second line contains five syllables.
3. The third line again contains seven syllables.

V. Drawing

Make a drawing of a natural object, using words to describe it (a flower blooming: green, growing, sweet, etc.) as the outline.

VI. Animal Charades

A. One person acts out an animal while others try to guess what is being imitated. Encourage students to choose animals that are a challenge. They also should know about them.
B. If students are having trouble, coach them with questions such as: "How does this animal eat? How does it react to humans?"
C. The student who guesses the animal gets the next turn.

VII. Through the Magic Window

Magic windows can be aimed at whatever students choose. They can make windows with their hands by matching thumbs and forefingers, investigate what they see, and share their findings.

VIII. Point of View

Look at an object from several points of view (i.e., a tree from: fifty feet away, five feet away, lying on the ground looking up, etc.). Discuss or have students write descriptions of the feelings they experienced from observing the different points of view.

IX. You Name It

A. The purpose of You Name It is to stimulate observation and imagination. Tell students to pretend they are Adam or Eve and are they are the first to see this object, getting to name it.
B. Tack a large sheet of paper to a board and attach a pencil to it with a string. Everyday place a natural object (weed, flower, osage orange, etc.) by the board. Encourage students to make up a name for the object and write it on the sheet of paper.

X. Reaction Board

A. This can be a large sheet of paper on a bulletin board which has an open-ended question. Encour-age students to write down their reactions to the question.
B. Some typical questions may be:

1. What do you like best about fall, winter, spring, or summer?
2. If you could be any animal, what would you be? Why?
3. If you could be alone anywhere outdoors, where would you go? Why?

C. This is a voluntary activity which gives students an opportunity to express themselves without worrying about being right or wrong.

Forest Manor

Objectives:

1. To help students recognize natural change in the environment.
2. To help students recognize man-made change in the environment.
3. To help students experience decision-making concerning land use.
4. To help students understand the advantages and disadvantages of land development.

Equipment:

1. Role-playing cards (Suggested roles are listed below.)

Unemployed worker
Farmer
Children
Factory owner
Construction company president
Merchant

Water	Deer
Air	Rabbit
Soil	Bird
Trees	Bank president
Mayor	Sportsman
Gas station owner	

2. Map of the area under consideration

I. Typical Class

A. *Idea of Change* — Have students go outside and look for two things in the environment: 1) Something that changes, 2) Something that stays exactly the same. Discuss their findings. The inevitable result is that nothing stays the same if left alone in the environment. Continue the discussion of change by asking questions concerning seasonal change in the environment. Finally, ask this question: "Do people and other things change the environment?'' Expect no answer; instead, play the game to dramatize it.

B. *The Game* — A staff member now assumes the role of a rich real estate developer who proposes to build a high-rise condominium in a wilderness area of the center. He describes, as vividly as possible, its special features such as gymnasiums, swimming pools, boating, etc. Students usually take offense to a condominium in a wilderness area. After a short discussion, take a vote to see who is for or against the building. The vote is usually negative.

C. *Role-playing* — Hand out role cards and explain to students that they should simulate the feelings and beliefs of the persons described on their role cards. Begin discussion of the condominium by playing the roles. After it has reached a peak and all have said a few words on their behalf, a chance is given to lobby for more support. Allow a few minutes for this.

D. *The Vote* — The leader now compiles a vote by asking each student to vote for or against the building. A reason also must be given for the vote. Invariably it is anti-condominium. Next, analyze the vote. Take away all of those roles which cannot vote such as air, water, soil, children, animals, trees, etc., and take a recount. This time the vote usually is pro-condominiums.

E. *The Aftermath* — This final discussion should consist of why nature cannot speak for itself, what students could do to protect or fight for a piece of land, and what local issues or examples are similar to this situation.

II. General Notes

Since role-playing is so important, the leader who plays the real estate developer also is extremely important. The more acting on the leader's part, the better because students respond in the same manner. If there is one who particularly understands what is going on, assign this student the role as mayor of the local community. Also, let this student help lead the discussion.

Trees and Such

chapter 4

Tree-mendous (Leaf Class)

Objectives:

1. To make students aware of similarities and differences in trees and their leaves.
2. To help students use ''process of elimination'' questions to identify specific trees and their leaves.
3. To help students understand the terms *simple, compound, toothed, lobed, alternate,* and *opposite.*
4. To help students appreciate the uniqueness and importance of trees and their leaves.

Equipment:

1. Trees of various types included on the Tree Key marked with yarn
2. One copy of the tree key for each group, possibly laminated to a piece of cardboard to increase durability
3. Hard writing surfaces
4. Pencils
5. Four leaves from each of five different trees

I. Introduction

A. Students meet in the Tree Encounter area. The instructor informs them that by the end of class, they can amaze their friends by identifying quite a number of trees and their leaves with one magical sheet of paper.
B. The importance of trees is discussed.

   1. What do trees do for us?
   2. What do they do for animals, mammals, and insects?
   3. How do they affect the land and weather?
   4. What would change if all the trees in this area were cut down?

C. Discuss some differences and similarities seen in trees.

II. Leaf Activities

A. Each student collects four leaves — two that are different and two that are similar — and explains how they are different and alike.

B. Students place their leaves on the ground in a sequence: from big to small, smooth-edged to deep lopes, rough surfaces to smooth, dark to light, or any other characteristics. This helps students become more aware of the slight variations in leaves.
C. Students pick and closely study leaves. Afterward they sit in a circle with their eyes closed and pass the leaves, one at a time, around the circle. Students should try to determine, by touch only, when their leaves have returned to them.

III. Definitions for using the Leaf Key (See Figure 4.1.)

A. *Compound* means that leaflets grow out of one stem to form a leaf. If individual leaflets do not have a large bud area at the base, they are part of a compound leaf. A compound leaf consists of three or more blades attached to a common stalk.
B. A *simple* leaf is a single blade growing directly out of the twig of the tree.
C. *Opposite* and *alternate* refer to the branching of twigs and branches. Alternate means that the twigs grow out of the stem with no other twig opposite it on the stem. Opposite branches grow opposite one another out of opposite sides of the stem. (See Figure 4.2.)
D. *Toothed* and *lobed* refer to the type of edges. Edges of leaves are either completely smooth, contain sharp teeth, or have larger variations called lobes.
E. *Even* and *uneven base* refer to the base of the leaf where each side meets the stem. If one side meets the stem at a higher place than the other, it has an uneven base.

IV. Students can begin to use the Leaf Key (See Figure 4.1.)

A. This Leaf Key identifies nine common trees' leaves, although many more types of leaves exist. If desired, a more complex leaf key can be devised to include them.
B. To identify a leaf, start with question 1. If the question is true, proceed to question 2. If question 1 is not true, proceed to the second question 1. Continue this procedure until the name of the leaf is discovered.

Leaf Key for Nine Common Trees

Figure 4.1

1. If the tree has needles, go to 2.
1. If the tree has leaves, go to 3.

2. If the needles come out from the branch in clusters of two, it is a red pine.
2. If the needles come out in clusters of five, it is a white pine.

3. If the leaves are compound, go to 4.
3. If the leaves are simple, go to 5.

4. If the leaves are opposite, it is an ash.
4. If the leaves are alternate, it is a hickory.

5. If the leaves are alternate, go to 6.
5. If the leaves are opposite, go to 8.

6. If the leaf has lobes but no teeth, it is an oak.
6. If the leaf has teeth, go to 7.

7. If the leaf margin is double-toothed, the base of the leave is uneven, and the leaf feels sandy, it is an elm.
7. If the teeth are all the same size, the base of the leaf is even, and the leaf feels smooth, it is a willow.

8. If the leaf has no teeth and no lobes, it is a dogwood.
8. If the leaf has lobes but no teeth, it is a maple.

Figure 4.2

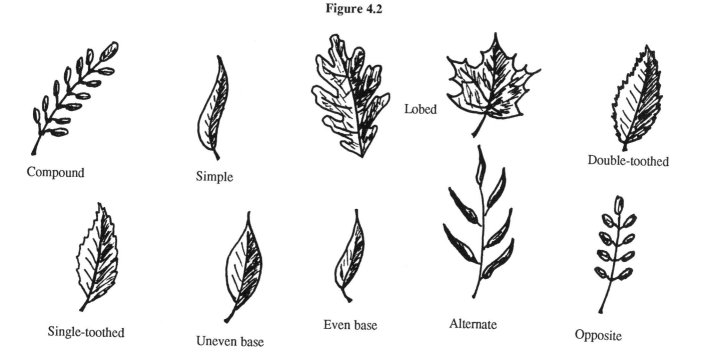

Compound Simple Lobed Double-toothed

Single-toothed Uneven base Even base Alternate Opposite

C. Have the class, as a group, pick a tree and use the key to demonstrate its use. Then split the class into three different groups and send each group to a different, previously marked area. In each area they should find six trees marked with yarn. Using the key, they should try to identify all six. Afterward, regroup and check the results.

V. Conclusion

A. Students find and stand by leaves that remind them of themselves and choose one word to describe their similarities. They take turns sharing each word or simply shout them all at once on the count of three.

Forestry

Objectives:

1. To help students understand that trees are a renewable natural resource.
2. To help students use simple forestry instruments to discover how one natural resource is managed on a *sustained yield* basis.
3. To help students understand that our forests have a multitude of uses (fuel, paper, food for man and wildlife, habitats, recreation, etc.).

Students first discuss the North American forest from the arrival of the first settlers to present. Next, take forest measurements to determine basal area, age of trees, and tree diameters. Based on this data, students mark a stand of trees for thinning. Finally, various uses of these trees are discussed and the *sustained yield* concept is brought into focus.

Equipment:

1. Eight cruise sticks (These are similar to yardsticks, having even marks approximately an inch apart — but not necessarily representing an inch. These marks are merely relative to compare one tree's width to another. The actual measuring unit is immaterial.)
2. Eight tree counters (any small, 1-inch wide stick or ice cream spoon)
3. Eight clipboards and pencils
4. Eight summary sheets for recording data
5. One large summary sheet used as a visual aid to explain the other sheets
6. About 120 pieces of yarn approximately 3 feet in length

I. Introduction

When the first settlers came to the New World, they found a seemingly unlimited forest. There were trees over 200 feet tall, many with diameters of over 25 feet. At this time, they saw no need to manage this tremendous, renewable resource.

Today, that picture of the past has long since been shattered. Most of the natural forest of North America has been cut completely — not only once, but in most cases four or five times. The demand on these forests has been increasing each year; and as that demand increases, the strain on these forests only grows heavier. We must learn to manage them so they remain for years to come.

Plants need the sun, water, air, and nutrients from the soil to grow tall and strong. If too many plants are growing on a plot of land, not all of them can get their basic requirements for survival. Some of the plants eventually assume a dominant role while others die. In the meantime, the fight for survival means that all the trees grow more slowly.

One job of the forest manager is thinning out the slower-growing trees to promote the growth of tall, healthy, fast-growing trees. By thinning a stand of timber and removing the undesirable group of trees, the growth potential of a forest is concentrated on fewer stronger trees, rather than being distributed over many slower-growing trees.

II. To the Instructor

This class is written for measurements in a particular stand of white pines planted in rows. Any stand of trees would work if the trees are of a fairly uniform species and age. The given constant of a 60-square-foot basal area is true only of white pines. That exact number is rather arbitrary and, if used for other species, would not make a difference to the ultimate outcome of the class.

A. Point Sampling

To find out how many trees, if any, need to be thinned out of an area, it is necessary to determine the basal area of the stand, and compare it to the optimal basal area a piece of land can maintain. (Basal area is the total area occupied by the tree stumps or bases in a stand.)

Since it is impractical to measure every tree's base area, an estimation method has been devised. Only a few trees representing the stand are chosen and measured. These must be randomly selected as typical of the stand. This random selection is called *point sampling.*

1. The trees are selected by using an *angle gauge*, which students can refer to as a tree counter. It has a fixed angle of view, defined by placing the 1-inch tree counter at arm's length (approximately 33 feet) from the user's eye. The tree counter selects trees for measuring.
2. The user stands in a randomly selected spot in the stand of trees and holds the tree counter at arm's length. With one eye closed and holding the tree counter in front of each tree within line of sight, the user then turns a complete circle. Any tree which fills out the tree-counter's 1-inch space or extends beyond its edges is a count tree and can be measured. By measuring only these trees, an estimate of the basal area of all the trees in the stand can be obtained.
3. A 1-inch angle gauge is called a "ten factor angle gauge." For the purposes of this class, this basal area factor of ten is a given. Therefore, the total number of count trees is multiplied by ten to obtain an estimate of the basal area in square feet per acre. Thus, twelve count trees have a basal area of 120 square feet per acre. In a stand of white pine, the best density for optimal growth is determined to be about 60 square feet per acre. If the basal area in a sample tree stand is higher than this, trees representing the total extra in basal area need to be thinned out. Thus, half of the trees

in a stand with a basal area of 120 square feet need to be thinned. It is important that students understand that this does not mean sixty trees are cut down, but a number of trees with a total basal area of 60 square feet.

4. The 360-degree circle made with the angle gauge is called a *plot* and the trees counted are those from one *plot*. If desired, select trees from more than one sampling point or plot and divide the total number of trees counted by the number of plots used. If a total of thirty-six trees are counted in three different plots, the average number of trees counted in each plot is twelve. This number multiplied by ten provides a basal area estimate of 120 square feet.

Total trees counted x basal area factor:

$$\frac{36 \times 10}{3 \text{ (number of plots)}} = \frac{360}{3} = 120 \text{ square feet/acre}$$

5. All trees located no further than thirty-three times (the distance in inches of the angle gauge from the eye) their diameter from the sample point are tallied. Therefore, a 1-inch diameter tree must be within 33 inches of the sampling point. If it is any father away, it is not a count tree. A 12-inch diameter tree can be recorded up to a distance of 12 x 33 inches, 396 inches, or 33 feet. Thus, for each full inch added to the base diameter of a tree, the tree can be 33 inches further away from the sampling point and still be tallied. This fact means that a fair sampling of trees of all sizes can be tallied.

6. To further illustrate this concept, it is convenient to presume that all trees are encircled with imaginary zones where radii are exactly thirty-three times the diameter of each tree. All these imaginary circles that overlap a given sample point represent trees to be tallied. Thus, the probability of tallying any given tree is proportional to its basal area. For the sighting angle discussed earlier, each tree tallied (regardless of its size or relative position to the sample point) represents a basal area of 10 square feet per acre.

B. Proof of Point Sample Method

Consider a 6-inch diameter tree.

1. Its imaginary plot radius is 16.5 feet, which is determined by:

$$\begin{array}{ccc} 6 \text{ inches} & \text{x} & 33 \text{ inches} \\ \text{(diameter)} & & \text{(sighted angle)} \end{array} = 198 \text{ inches or } 16.5 \text{ feet}$$

2. This hypothetical zone (16.5 feet) represents an imaginary plot of 0.0196 acre around each 6-inch diameter tree, which is determined by:

$$\frac{\pi r^2}{\text{square feet in acre}} = \frac{3.14 \times 16.5^2}{43,560} = 0.0196 \text{ acre}$$

3. To determine the estimated number of 6-inch trees per acre, divide 0.0196 into 1.00 acre, which results in 51.02. Thus, when a 6-inch diameter tree is tallied, it is assumed that there are 51.02 such trees per acre.

4. Basal area per tree in square feet for a 6-inch tree is determined by multiplying the constant, 0.005454, by the diameter in inches squared.

$$0.005454 \text{ d}^2 = \text{basal area (in square feet)}$$

Therefore, a 6-inch tree:

$$0.005454 \times 6^2 = 0.196 \text{ square feet}$$

5. To determine basal area factor, BA of a 6-inch tree (0.196 square feet) x 51.02 trees per acre yields a constant Basal Area Factor of 10 square feet

$$0.196 \times 51.02 = 9.99992 = 10 \text{ square feet}$$

III. The Class

The class is greatly simplified for students and does not deal with the theory and proof of point sampling since that is not an objective. However, the main objective is for students to understand there is a method for measuring a forest and for them to imitate that method to learn the strategies used by foresters.

A. Introduction

1. Do you notice anything strange about this stand of pines? (They are all in rows which shows they were planted.)

2. Because people planted them, we probably can assume that they are all the same age. How do you figure the age of a tree? (Cut it down and count the rings. With pine trees, however, simply count each whorl or circle of branches. The sections between each whorl is a year's growth.) Each student can pick out a tree and figure its age.

3. Now that we know these trees are all about the same age, what else is strange about them? (Some are fat and some are thin.)

4. What could cause that? (Some get more food, air, water, minerals, and sunlight than others.)

5. Everyone find two trees close together, one that is doing well and one that is not. Look at their tops as well as their bases. Which tree has the most green on it? (The fat one.) What would happen if we had enough food here for fifty people and we had a hundred people who

wanted to eat? (Everyone would get some if we shared, but no one would get full.) That may not be too bad for one meal, but what if it happened for a week? How would you feel? That is the way it is here. There is only enough food for a certain number of trees. None of them die right away, but none are really strong and healthy. Some trees are weak, starving, and skinny. Even the big ones are not doing as well as they could.

6. What might be the problem? (There are too many trees.)

7. What would be a logical course to follow if you were a forester trying to solve this problem? (Some should be thinned out.)

8. Would a forester just guess how many trees should be removed or use a more scientific method? Foresters work in cruising teams of three. Working in teams of three and using the same tools as a forester, we are going to figure out how to thin a forest.

B. Procedure

1. A team member stands in one spot and holds the tree counter at arm's length in front of him. (See Figure 4.3.) The member closes one eye as if to aim a gun. Every tree which sticks out around the tree counter's edge can be counted and measured. (See Figure 4.4.)

2. Another team member then goes to that count tree and measures the diameter with the cruise stick. When used for measuring, the cruise stick must be held at arm's length against the tree. The forester closes one eye and lines up the edge of the stick with the side of the tree. Holding the head steady, he sights along the stick to the opposite side of the tree and, thus, measures the tree by the corresponding mark on the cruise stick. (See Figure 4.5.) If they have not been explained by this point, the terms *diameter, circumference,* and *area* should be clearly defined.

3. The student with the cruise stick yells the measurement to the student in charge of the summary sheet, who then places a tally mark on the paper in the appropriate place. (See Figure 4.6.)

4. A plot is complete when the person with the tree counter goes all the way around in a 360-degree circle and every tree which sticks out beyond the edge of the tree counter is measured and tallied. Once students are taught how to use the instruments, several people can run through a mock plot. Students can then begin to work on their plots. They should try to make at least three plots and record all data on one summary sheet. As each new plot is started, the three students should change roles so each has a chance to experience the different roles. Once all three plots are finished, which takes about twenty minutes, bring the group together and analyze the data.

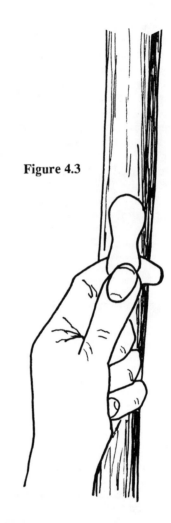

Figure 4.3

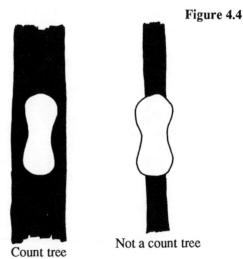

Figure 4.4

Count tree   Not a count tree   On the line (count every other

C. The Data

1. Add the total number of trees tallied and place that number in the proper blank on the summary sheet.

2. Multiply the number of trees tallied by ten.

3. To obtain an average per plot, divide this total by the number of plots or point samples taken.

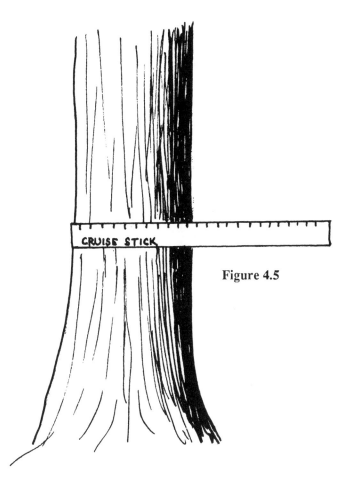

CRUISE STICK

Figure 4.5

FORESTRY CRUISE SUMMARY SHEET	
Diameter	Trees Tallied
6	ⅢⅢ ⅢⅠ
8	ⅢⅢ Ⅰ
10	ⅢⅢ Ⅱ
12	Ⅱ
14	Ⅱ
	Total trees tallied 26
	Total BA/acre 87 square feet
10 x Total trees tallied ÷ number of plots	

Figure 4.6

This final answer is the Basal Area (BA) per acre for that stand of white pines.

4. To obtain optimal growth, the goal for this stand of trees is 60 square feet. If the final answer is 120 square feet, then half the trees need to be removed. If the final answer is 90, for example, about a third of the trees need to be removed.

5. To explain basal area, show students a tree stump, which may be marked off in square inches. The total BA of the forest is all of the tree stumps areas added together. If all the trees were chopped down and the tops of the stumps were measured and added together, the figure would represent the total BA of the forest.
6. Have students figure how many trees must be removed for the forest to reach its fullest potential.
7. Which trees should be thinned? (Thin, crooked, dying, or diseased trees should be removed so the bigger ones with the best head start have space.)
8. Students then take pieces of yarn and quickly move through the forest, laying yarn around the trees to be cut.

IV. Conclusion

A. What are some uses of wood? (Furniture, houses, firewood, tools, paper.) Students' answers almost always deal only with man's needs. If they are told that a logging company has bought all the trees in this forest for the needs they just mentioned and the company is chopping all the trees down within a week, the students immediately become upset. They ask why the trees must all be chopped down.
B. What are some other uses for the trees? (They look nice. They make homes for wildlife, protect the soil, and are great places to explore and hike.)
C. So you are saying we should not cut down all these trees? Do you agree we need trees for many things, including wildlife as well as houses? Should we manage our trees the way the logging company wants or the way we just did when we marked only some of the trees? What we just did is called *sustained yield*, meaning the yield of the forest is sustained or kept the same from year to year if we correctly manage it. Each year we cut only what grows and no more so that we never run out of trees.
D. To gather the yarn from the trees, race to see who can bring back the most pieces.

Incredible Edibles

Objectives:

1. To make students aware of the tremendous variety of edible plants which surround them.
2. To help students recognize and taste certain edible plants.
3. To help students realize the values and benefits of plants previously thought to be unwanted weeds.
4. To make students aware that potential hazards in gathering and eating wild plants exist and they should not experiment with wild plants.

Although edible wild plants are available throughout the year, the greatest abundance and variety are found in fall or late spring.

I. Pre-experience Ideas

A. Experiment with taste sensitivity and areas of the tongue. Have students close their eyes and touch various tastes (sweet, sour, salty, bitter) to their tongues. Note which areas are more sensitive to certain flavors.

B. Experiment with the relationship of smell and taste. Have students close their eyes. Hold apple slices under their noses and let them taste pear slices. Ask what they tasted. Try other combinations of foods with similar textures. Have students close their eyes and hold their noses while tasting various foods. Why do tastes change when we have colds?

C. Study nutritional values of foods at home by examining labels. Plan a balanced menu using foods investigated. Note the amount of additives and preservatives in foods. Discuss why they are needed to process foods for our society, which is removed from farms, and compare them to the fresh foods eaten by pioneers and native Americans.

II. Introduction

A. Native Americans and pioneers could not run to the grocery store to get food. They had to rely on wild foods, their gardens and wild game. We are going to try some plants which were eaten by them. Who collected the plants? (Women and children.) Although many wild plants are good for you and taste good, many others are poisonous. How would a pioneer know what was safe to eat? Is it good to try a bit and see if you get sick? Some plants are so poisonous even a small amount makes you very sick or even kills you. (Ask someone who knows the plants of the area?) How can we find out today? (Ask someone or look plants up in a book.)

B. It is very important to follow a few guidelines when collecting plants.

1. Know exactly what plants are before eating them. Many poisonous plants resemble edible ones.
2. We cannot eat plants just because animals do.

   a. Squirrels eat all mushrooms.
   b. Horses eat poison ivy.
   c. Birds eat deadly nightshade.

3. Not all parts of plants may be edible.

   a. Rhubarb stalks are good, but the leaves are poisonous.
   b. Elderberries are good, but the rest of the plant is poisonous.
   c. Wild cherries are edible, but the seeds are poisonous.

4. Berries are not all edible.

   a. Blue or black are 80 percent edible.
   b. Red are 50 percent edible.
   c. White are less than 10 percent edible.

5. If something is hot, tastes bitter or like almonds, do not eat it.
6. When eating edible plants for the first time, eat only a small amount.

   a. You may be allergic to it.
   b. Wild plants are different than most foods and your stomach may not be ready for a great quantity. If there is no reaction the first time, you can eat a little more the next time.

III. Activities

A. Have students pretend to be pioneers who are new to this area and need to gather plants for a meal. Each student brings back one plant thought to be edible. After the group reassembles, have students show what they brought back and explain why they thought their plants looked edible. If you know plants are edible, have students taste them. If plants are poisonous or you are unsure, explain that you cannot take a chance. Before students go collecting, stress that no one should taste anything until the leader gives permission and they should only take a small part of the plant so it is harmed as little as possible.

B. Edible Hike

The leader takes the class on a hike through the area, pointing out edible and poisonous plants. The class can taste the edible plants.

C. Edible plants include:

1. Pines, white and red
2. Wild carrot called queen anne's lace
3. Nuts, hickory, hazelnuts, walnuts
4. Sassafras
5. Wild grapes (Compare to solomon's seal and virginia creeper.)
6. Sumac, smooth and staghorn (with the red fuzzy berries)
7. Chickory
8. Cattail
9. Burdock
10. Milkweed
11. Peppergrass
12. Sorrel and clover
13. Highbush cranberry
14. Wild strawberry
15. Jewelweed
16. Dandelions
17. Cherries
18. Elderberries
19. Rosehips

D. Poisonous plants include:

1. Poison ivy
2. Deadly nightshade
3. Solomon's seal

4. Holly
5. Horse chestnut and buckeye
6. Dogwood berries
7. Mushrooms and fungus

The leader checks for available plants in the area before class. At this time, any plants which are not in the immediate area that the leader wants to display also are collected. Only the leader should pick and pass out samples. Explain that animals in the area depend on these plants for food and we destroy their food supply by wasting them.

IV. Conclusion

Review the tasted plants. Ask which were favorites. If planning a meal, which could be the main course? The salad, drink, vegetable, or dessert? Could we all live off the land? Discuss the time and work needed. How much land would it take to support us? Once again stress caution in eating wild plants and review the rules.

V. Post-Experience Activities

A. Prepare and eat an entire meal of wild plants.
B. Take a hike and list edible or poisonous plants found around the home or school.
C. Investigate medical uses of plants.
D. Experiment by feeding different wild plants to classroom animals. Find out which ones are preferred by animals and compare the results to those preferred by humans.
E. Try growing wild seeds or starting an edible plant garden.

Succession Study

This class was developed for a particular lake; however, it can be altered to fit the features of any lake.

Objectives:

1. To help students comprehend the idea of succession through on-site analysis and evaluation.
2. To help students see their roles in hindering or fostering succession in city or country.
3. To help students utilize the scientific method (gathering information, interpreting it) through self-directed activities.
4. To help students comprehend how and why the lake is filling in.

5. To help students use an interdisciplinary approach (science, math, social studies, creative expression) to gather information on succession.

I. Introduction

A mystery is to be solved by a group of private detectives who are gathered in front of the instructor. It seems that the lake has been disappearing. This group has been hired to determine the exact details surrounding its disappearance.

II. At the Site

Questions are asked on the way to the site or when the group arrives at the lake edge.

A. Where is the original lakeshore? How do we know? (A ridge circles the lake about 75 yards from the lakeshore and shows that the original lakeshore was formed by glacial till and the lake's wave action.)
B. Is the lake man-made or natural? How was the lake formed?

1. Usually cattails and reeds extending into a lake, an irregular shoreline, lack of cottages, levees, or build-ups are evidence of a natural lake. However, a man-made lake could acquire a natural look over a period of years. Look for dams, inlet streams, artesian wells, and natural lakeshore ridges. This lake appears to be natural. Have students tell if they have ever visited a man-made lake.
2. Ten thousand years ago glaciers, which were more than two miles high, scored Michigan and left a flat plain. (See Figure 4.7.) In Michigan glaciers did not gouge out lakes. Instead, when glaciers melted due to warmer climates and receded north, large chunks broke off amid soil and rock (alluvial materials) that glaciers accumulated in their journey south. (See Figure 4.8.) When these glacial chunks melted, a kettle lake resulted. (See Figure 4.9.)

Figure 4.7

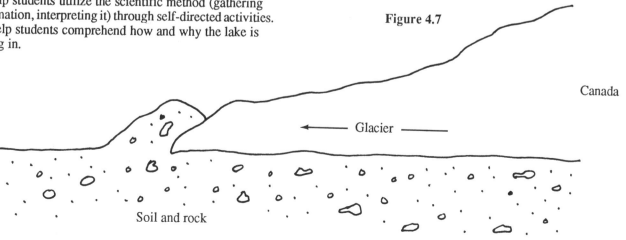

Canada

Glacier

Soil and rock

Figure 4.8

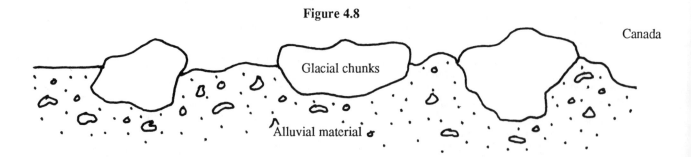

Canada

Glacial chunks

Alluvial material

C. How is the lake filling in? (Plant growth into the lake, erosion, decomposition of dead plants and animals, and wave action are all factors contributing to the lake's filling in.)

D. Why are no large trees by the lakeshore? (Answer this question later because it helps create the detective mood.)

III. Six Measurements to Gather Information

Split students into three groups to take certain measurements in four different areas, interpret their findings, and record these findings on data sheets. (See Figure 4.15).

Selection of the four areas is a crucial aspect of this class. They must be contrasting enough to clearly show succession. If not, students do not grasp the concept as well. If possible, use paths, beginnings of tree sectors, or brush sectors as dividing lines for the four areas which correspond roughly to stages in the natural development (reeds to dogwood to elm to oak/hickory forest).

The climax forest farthest from the lake is designated as Area 4. The marshy section near the lakeshore is Area 1. Areas 2 and 3 are between the other two stages.

A. Dominant Plant

Have each group decide the two dominant plants of each area and write names for them in the proper space on the data sheet. If necessary, help students define *dominant,* which means the most common plant in the area or the plant with the greatest spatial reach that seems to be taking over. For example, many violets may be in an oak forest, but an oak would be considered the dominant plant in that area. Naturally, students could consider both oaks and violets as dominant in an area.

Next, have students invent descriptive names for the two selected plants. This helps students become involved in class and sharpens their eyesight and descriptive powers. It is not necessary that students correctly identify plants; that is not the primary reason for this class. Of course, a good teachable moment, involving a plant name and a bit of the plants background, may enhance the class.

B. Plant Circumference

Many students may not know this word, so this is a good time to add it to their vocabulary. Have students measure how far around the plant is. Tape measures could be used; or new units could be made up such as number of pencils around, hands around, or fingers around. The importance of measurements lies in their relationships, not in their exactness. Students may measure and say, "This plant stem is about as big around as an elephant's leg . . . a giraffe's neck . . . a drinking straw."

C. Plant Height

Figure 4.9

Canada

Kettle lakes

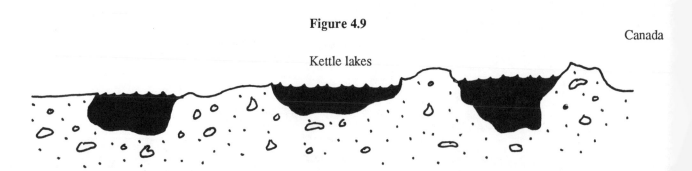

Students measure height in the same way as circumference: "This plant is about as tall as a cabin . . . three-story apartment . . . three gorillas . . ." Another good idea is to have students use their own heights and measuring tools. Have a meter stick handy, measure a student of average height (not the shortest nor tallest in the group), and use that measurement to gauge plant height. As for oaks or cherry trees, use the artist's method of figuring height.

Figure 4.10

1. Select one student who is about 5 feet tall. For shorter reeds and plants, estimate plant heights in relation to this person. For example, a plant one-half as tall would equal 2 1/2 feet). Metric measurements also may be used in this exercise.
2. For larger trees and shrubs:

   a. Have a person who is 5 feet tall stand in front of a tree.
   b. With arm extended, back up until the size of your thumb appears as tall as the person standing against the tree. (See Figure 4.10.)
   c. Your thumb is now a ruler. Move it up the tree, and count the number of thumbs in the tree's height. Multiply this number by 5 feet to get the approximate tree height.

D. Moisture

   1. The amount of moisture a plant receives is an important index of plant growth. Too much or too little water in the soil means nothing will grow.
   2. Have students guess the amount of moisture an area receives and retains: a lot, some, or a little. This is an unscientific method, but students can easily figure it out. Gathering pinches of soil, observing standing water, noting proximity of an area to a lake or muddy conditions are all clues students use in this experiment.
   3. If you have time and want to be more scientific, get a No. 10 can with both ends cut off and stomp it about halfway into the ground. Fill the can with water to a certain level and time how long it takes the soil to absorb the water. This shows how permeable the soil is, whether it can hold water well or not. Comparisons of the four areas shed some light on why certain plants are not growing in certain areas.

E. Natural Litter

Ask students what litter is (garbage, cans, paper, etc.). Then ask what natural litter is (leaves, grass, dead plants, etc.). Have students estimate the amount in each area: a lot, some, or a little.

F. Light

This is a simple experiment. Bring a piece of white paper for each group. Hold a hand or pencil over the paper. If there is a very distinct shadow, there is a good deal of light in the area. If the shadow is less distinct, there is a smaller amount of light. Make allowances for cloudiness.

IV. **Interpreting Findings Back in the Classroom**

A. Have students list the basic needs of plants:

   1. Water
   2. Rich soil
   3. Carbon dioxide
   4. Growing room
   5. Light

This information comes in handy later. On a chalkboard make a chart similar to the ones used by students. Have spokespersons from each group read the results of their research. Afterward, summarize the data from all the groups onto the main chart.

B. Probable Findings

Going from lake to oak/hickory forest:

   1. Moisture decreases.
   2. Amount of light decreases.
   3. Amount of natural litter increases.
   4. Circumference and height of plants increases.

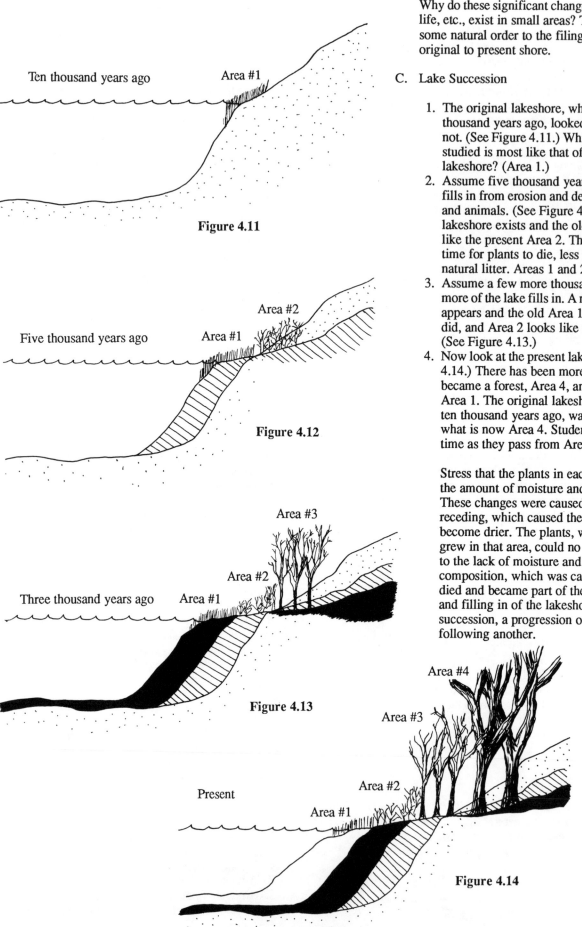

Figure 4.11

Figure 4.12

Figure 4.13

Figure 4.14

Why do these significant changes in moisture, plant life, etc., exist in small areas? There seems to be some natural order to the filing in of the lake from original to present shore.

C. Lake Succession

1. The original lakeshore, which existed ten thousand years ago, looked much like it does not. (See Figure 4.11.) Which of the four areas studied is most like that of the original lakeshore? (Area 1.)

2. Assume five thousand years pass and the lake fills in from erosion and decaying water plants and animals. (See Figure 4.12.) A new lakeshore exists and the old lakeshore looks like the present Area 2. There has been more time for plants to die, less moisture, and more natural litter. Areas 1 and 2 are filling in.

3. Assume a few more thousand years pass and more of the lake fills in. A new lakeshore again appears and the old Area 1 looks like Area 2 did, and Area 2 looks like the present Area 3. (See Figure 4.13.)

4. Now look at the present lakeshore. (See Figure 4.14.) There has been more erosion. Area 3 became a forest, Area 4, and there is a new Area 1. The original lakeshore, which existed ten thousand years ago, was at the edge of what is now Area 4. Students walk through time as they pass from Area 1 to Area 4.

Stress that the plants in each area changed as the amount of moisture and sunlight changed. These changes were caused by the lakeshore receding, which caused the adjacent areas to become drier. The plants, which originally grew in that area, could no longer survive due to the lack of moisture and change in the soil composition, which was caused as the plants died and became part of the soil. The erosion and filling in of the lakeshore causes natural succession, a progression of plant types, one following another.

V. Follow-up Questions

A. What happens to the lake if succession continues without man stopping it? (It would probably fill in. If it continues, whole areas could become an oak/hickory forest. Tell students that the succession cannot go any further because the oak/hickory forest is the climax in this area.

B. Why isn't the athletic field or dining hall area a forest? (Farmers cut down trees to make fields for crops. Note the barb wire grooves along the boundaries of the oak forest in some areas. This shows a boundary between the field and forest that would not have been there without the farmer's help.)

C. Have you ever stopped succession? (Sure. For example, mowing a lawn, weeding an area, or occasionally sledding down a hill in winter stops succession. Suggest students tell their parents that they are going to go stop succession the next time they mow the lawn.

VI. Follow-up Activities

A. Find an area near school that has a similar example of natural development (a vacant lot growing back, a forest area encroaching upon the school yard). Teachers might want to keep a certain part of the school yard unmowed so they can chart differences between it and a mowed area.

B. Apply the concept of succession to the city. A plot first may be a farmer's field (or a vacant lot), then scattered houses, then a subdivision, etc.

LAKE SUCCESSION DATA SHEET

Area	Dominant Plants	Height	Circumference	Moisture	Natural Litter	Light
#1	1. 2.	1. 2.	1. 2.	Little Some Much	Little Some Much	Little Some Much
#2	1. 2.	1. 2.	1. 2.	Little Some Much	Little Some Much	Little Some Much
#3	1. 2.	1. 2.	1. 2.	Little Some Much	Little Some Much	Little Some Much
#4	1. 2.	1. 2.	1. 2.	Little Some Much	Little Some Much	Little Some Much

Figure 4.15

Back in Time

chapter 5

Pioneer Crafts Fair

Objectives:

1. To teach students some of the historical facts concerning early American life.
2. To help students experience some of the daily activities of early Americans.
3. To give students a chance to work independently on one or more personal projects which they can take home.

Structure:

The entire crafts fair lasts about two hours. Students meet in a central location and are given an introduction to the class. Afterward they split into four groups. Each group then meets as a mini-class at a different location where one activity is demonstrated and explained. At the end of fifteen minutes, each group rotates to the next mini-class. After an hour, each group has visited four demonstrations and received samples of some early American crafts. These four demonstrations could include spinning wool, natural dyeing, soapmaking, butter churning, candlemaking, pioneer cooking, tools, weaving, or other appropriate crafts.

After an hour of rotating to different demonstrations, the whole group gathers again in the central location. They then are informed about the activities available for the next hour or so. Students may go to any activity, stay as long as they are interested, and then go to another activity. Eight to ten activities are ideal. In addition to the aforementioned crafts, demonstrations may include weaving, edible wild plants, making preserves or apple butter, Dutch oven cooking, making pioneer toys or decorations, cornhusk dolls, ice cream, tapping maple trees and making maple syrup, and many others. Students can make either an item to take home or something to eat at each activity.

Butter Churning

Equipment

1. Whipping cream
2. Salt
3. Water
4. Bowl
5. Butter churn
6. Spatula
7. Small glass jars

I. Procedure

A. Pour whipping cream into a churn. Baby food jars may be used in place of a churn since a small amount of cream turns to butter in five to ten minutes. Fill the jars about half full.
B. Begin churning or shaking with a steady rhythm.
C. Butter finishes in approximately twenty to thirty minutes. Butter, however, cannot be timed; it must be checked from time to time by looking in the churn to see if it is completed.
D. Butter is finished when the whipping cream changes to a rough chunky consistency with a yellowish tinge. A watery, milky substance also surrounds the yellow butter chunks.
E. When butter finishes, pour the contents from the churn into a bowl. Gather the butter into one lump by using a spatula and pour the milky liquid out.
F. Wash the butter by pouring cool, clear water over it into a bowl and mashing the butter with a spatula. When water turns milky, pour it out. Use several changes of water until the water comes out clear.
G. Add salt according to taste and enjoy the butter on a cracker.

II. Butter Tidbits

A. Cream is obtained from cow's milk. After a cow is milked, the milk sets until the cream rises to the top. Afterward, the cream is skimmed off and skimmed milk is left behind. Why doesn't cream rise to the top on our milk?
B. Buttermilk, the milky substance surrounding the butter chunks, is left over from making butter.
C. Washing the butter is the process which rids the butter of the sour buttermilk taste.
D. Steady rhythm is necessary in churning because the butter comes faster and churning is not as tiring. A song used to keep time (rhythm) while churning:

> *"Come, butter come. Come butter, come.*
> *Johnny's waiting at the gate.*
> *Waiting for his buttercake.*
> *Come, butter come. Come butter, come."*

This song can be sung in a round.

E. The natural yellow coloring in butter is a chlorophyll substance, which is obtained as the cow eats green grass containing chlorophyll.

F. Pioneers made butter only as time and cream allowed; it was not an every-day process. Often by the time they had collected enough, the cream had gone sour; therefore, sour cream butter was very common. Salt also was quite a luxury and was not always obtainable.

G. Pioneer women, while moving westward, simply filled a covered container with cream and hung it on the back of their jostling wagons. By the time they stopped for the night, the butter was ready.

Spinning

Equipment:

1. One or more sets of wool cards
2. One or more drop spindles
3. A spinning wheel, if possible
4. Uncleaned, unspun raw wool
5. Knitted or woven samples of dyed or undyed homespun wool

I. Fibers for Spinning

A. Cotton, grown in warm climates, and wool are spun into yarn in much the same way.

B. Flax, a long-stemmed plant, is soaked in water and pulled through nails in a board. Afterward, the long fibers which remain are spun into yarn. Linen is the material woven from this yarn. Flax fibers mixed with wool were used to produce a material known as ''linsey-woolsey,'' which was commonly used for shirts and underwear.

C. Many other hairs, furs, and fibers can be spun into yarn. These may include buffalo wool, dog hair, goat hair, llama wool, angora, and rabbit fur.

II. Advantages of Wool

A. Wool is water-resistant and warmer than most other fibers; it provides more insulation when it is wet than other materials.

B. Leather, used for garments by frontier Americans, is wind-proof and very tough. However it is neither very warm (unless the fur is left on) nor very water-resistant. Also, an animal must be killed to obtain leather, while wool can be obtained numerous times from one animal.

C. Wool was easier for the frontier people to obtain since they could raise their own sheep. Cotton and flax, grown in the south, was quite expensive for people in the north.

III. Wool

A. Sheep can be sheared twice a year, fall and spring. Shearing once, only in the spring, yields a longer fiber, which is better for hand-spinning. Years ago one sheep might give two to six pounds of wool.

Now with special breeding, eight to twelve pounds is common. Ten pounds of wool is enough to make five or more sweaters and about twenty to thirty stocking caps.

B. The curly nature of wool fibers allows them to bind together when spun. The longer the fiber, the easier it is to spin and the stronger the resulting yarn.

C. The natural oil in wool is called lanolin, which gives wool its water-resistant nature. Spinning wool before washing is called spinning in the grease, which is easier to do than spinning wool after much of the lanolin has been washed out.

D. The air spaces between the curly wool fibers and the strands of yarn in a garment give wool its insulating properties.

IV. Preparing for Spinning

A. Usually the raw wool is hand-picked first to remove burrs, pieces of straw, and other large pieces of debris.

B. The wool then can be dyed. The dyeing process, particularly if the wool is boiled, removes much of the lanolin. This makes the wool dry, brittle, and difficult to spin. Therefore, dyeing is usually the very last step in preparing the yarn for use.

C. Next, the wool is carded with large, wooden brushes having small wire teeth. This straightens the fibers, placing them parallel to one another, and removes more dirt and debris.

    1. Place a small handful of wool in the middle of one card. Holding the other card on top of the first, with the handle pointing in the opposite direction, pull the cards apart, stretching the fibers. Continue this combing until the fibers appear smooth and parallel.

    2. To remove the wool, turn one card around so the handles of both cards are pointing in the same direction. Lightly push the cards back and forth, rolling the wool. The wool rolls out near the bottom of one of the cards. Peel it off in a long roll, which is called a *rolag*. It is now ready for spinning.

V. The Drop Spindle

A. The drop spindle is an ancient device, which has been used for spinning fibers since biblical times. Used across the ancient world and into the modern world by primitive cultures, it is still used in parts of the world today.

B. A drop spindle is simply a 10- to 12-inch dowel with a knob of wood at the end. (See Figure 5.1.) It even can be made by sticking a knitting needle through a potato. American Indians used a longer stick and, bracing one end on the ground, twirled the dowel against their thighs with their hands.

    1. A starter yarn of 18 inches or so needs to be placed on the spindle. It is tied to the dowel just above the knob, catching around the end

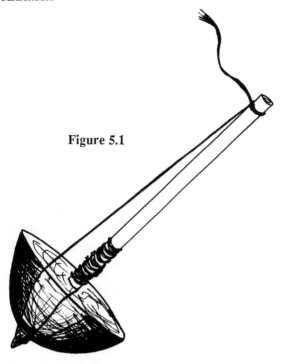

Figure 5.1

of the dowel which protrudes from the bottom of the knob. Then it is brought up to the top of the dowel, which is notched, and fastened with a half-hitch.

2. The rolag to be spun is stretched out somewhat thinly so about 6 inches of fuzzy wool hangs down. This piece should be pinched and overlapped against the fuzzy end of the starter yarn. Holding the overlapping pieces in one hand, give the top of the dowel a spin between the fingers as one would spin a top. This should fasten the new piece of wool to the end of the starter yarn with enough strength to support the weight of the dowel; thus both hands are freed.

3. Continue to alternately spin the dowel and stretch out the rolag so a section of unspun fuzzy wool is continually getting caught in the spin. As the wool spins, let the spindle drop so the yarn continually increases in length. When the spindle is ready to hit the floor, stop, remove the yarn from both ends of the dowel, and wind it onto the dowel just above the knob. Wind all but 12 to 15 inches, which again is anchored on the dowel. Continue this process, adding rolags as needed.

4. When the dowel fills with yarn, remove the yarn and roll it into a ball or wind it around the elbow and hand, making a loop. Afterward, tie it in several places to prevent tangling. It can be washed and dyed in this form. A *niddy-noddy*, special wooden stick which is shaped like a capital I and 18 inches long, was used to make the loops. One full time around the four tips of the I measured off 2 yards of yarn; forty times around was a standard measurement for a skein of yarn, which was then 80 yards.

VI. The Spinning Wheel

A. The spinning wheel, invented during the eighteenth century in England, was a much faster method of spinning because the spinner's foot turned a wheel that turned the spindle. The flywheel wound the yarn, so the spinner never had to stop and wind the yarn onto the spindle. This meant a spinner could spin through a pile of rolags without ever stopping the spinning process.

B. The spinner's hands still operate in the same manner as with the drop spindle, pulling out lengths of fuzzy unspun wool and letting them slide through the fingers to be caught up in the spin. The spun wool is held lightly and the wheel gently draws it through the fingers, through the flywheel, and onto the spindle.

C. Early Americans, moving to the frontier, may or may not have taken their spinning wheels. Wheels made entirely from wood, had leather hinges on the moving parts. Thus, they could be made once pioneers arrived at their homesteads. In the meantime, the dependable, ancient drop spindles were in service.

Candlemaking

Equipment:

1. Paraffin
2. Crayons
3. String or wick
4. Double boiler
5. Water

Pioneers were up at dawn, busy with their chores, because no precious hour of daylight could be wasted. At night the glow of the fire often provided the only light in the room. Candles were used very sparingly since candlemaking was a difficult job. Those who could afford them bought candles and soap in the chandler's shop. Chandlers also traveled from house to house making candles and were not welcome neighbors because candlemaking was a greasy, smelly trade. Since the materials, equipment, and parts of the processes were similar, they also made soap.

The lighting problem often was solved by burning fat or wax, which soaked or surrounded some kind of wick. Tallow for candles was hard animal fat, which came chiefly from cattle and sheep. A bull yielded enough fat to make 26 dozen candles.

I. The Chandler

The chandler did not merely melt fat; he *tried* or *rendered* it by boiling so the fibrous material in it rose to the surface. Afterward, it was skimmed off. If the chandler was dipping candles, he would ladle the hot fat into a rectangular vat and allow it to cool some. Very hot tallow left a thick deposit on a dipped wick or, worse, it melted the accumulation of earlier dippings. Before 1750, chandlers made their best candles from beeswax or bayberry wax, which was obtained by boiling berries from the beach shrub.

II. In the Home

A housewife never threw away any fat; she rendered and stored it in pottery crocks. In very early days, the family burned it in grease lamps for light; however, it was smelly and smoky. Thus, candles were better. In the home, making candles was a job for women and girls. Several long wicks made of rolled cotton, spun flax, silky down from milkweeds, or other string-type

I. Introduction

A. White wool, especially without soap, cannot be kept clean.
B. Having bright clothes became a sign of wealth and power.

 1. Purple, for royalty, came from shellfish.
 2. Madder root, the red as in Oriental rugs, was

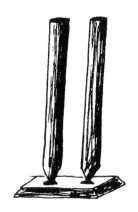

Figure 5.2

materials, were hung over a stick and dipped in melted tallow, so numerous candles were made at once. (See Figure 5.2.) Dipping continued until the candles had become thick. Later, tin molds were used; melted tallow was poured into the molds and allowed to cool around the homemade wicks. As many as six, eight, or more candles could be made at once.

mixed with ox blood, cow manure, rancid olive oil, galls, and alum.
 3. Indigo, the original dye for blue jeans, came from a plant.
 4. Cochineal, insects from Mexico, made bright red.

C. Native Americans who lived in the Southwest dyed baskets with colors that symbolized various things.

III. Today

Cheap candles are now made of paraffin, a petroleum product unknown to our forefathers. Their cheap candles were made of tallow until the early years of the twentieth century.

V. Procedure

1. Melt wax and crayons (for color, if desired) in double boiler.
2. Cut string for wicks.
3. Dip the wicks in wax, then in cold water. Dipping candles in cold water speeds up hardening. Pioneers did not dip candles in water because the layers of wax followed by layers of water made the candles flaky.
4. Repeat this procedure, alternating dips of wax and water, until candles are the desired thickness.
5. Candles can be molded into different shapes while they are warm. They also may be shaped and redipped.

Natural Dyeing

Equipment:

1. Twigs to scrape
2. Several prepared dye baths in No. 10 cans on a fire
3. Sample plants

II. Collecting Material

A. Color is determined by time of year, location of plant, amount of dye material, and amount of water. Use good conservation practices in collecting.
B. Have sample to show color and subtle differences. Also have samples of plants.
C. A few plants might include:

 1. Rose hips — light pink
 2. Walnut hulls — deep toasty brown
 3. Burdock roots — deep gray
 4. Buckthorn bark — yellow
 5. Sumac berries — tan
 6. Black oak bark — bright yellow
 7. Mullein — soft yellow
 8. Sassafras bark — rose tan
 9. Cherry bark — butter yellow
 10. Onion skins — bright gold
 11. Cochineal — purple to rosy red

III. Preparing the Dyebath

A. Scrape to get dye from the bark, wash and cut up roots, and dry marigold flowers. (The group can scrape twigs.)

B. Soak overnight — roots and bark especially.
C. Boil. Time depends on the part of the plant. For example, bark should boil longer than leaves. (Show dyes on fire boiling.)
D. Strain the dyebath through a screen. (The group may help with this.)

IV. Mordanting of Material

A. The mordant keeps the color from fading. Some dyes, such as sumac and walnut, contain enough natural mordant that no more is needed.
B. Add mordant to boiling water, using one teaspoon of alum per gallon.
C. Add the wool to the alum water and simmer for a few minutes.
D. Try not to shock the wool by allowing sudden temperature changes which weaken fibers.
E. The type of mordant affects color (e.g., alum brings out yellows).
F. Other mordants include:

    1. Drip lye — especially from cedar ashes which have been boiled and strained
    2. Chamber lye — urine
    3. Salt, soda, vinegar, ammonia
    4. Metals — iron, brass, tin (Use kettles made from that material.)
    5. Alum — from plant roots or mineral deposits

G. When mordant is added to the dyebath, wool is boiled only once.

V. Dyeing

A. Natural fibers must be used: wool, cotton, flax, silk.
B. Wool does not have to be spun first; however, it does take dye better if it is washed first.
C. Add wet wool to dyebath. (The group may add the prepared wool to dyes they have strained.)
D. Simmer for five minutes to one hour, depending on darkness of color desired. Do not boil. (As the group adds wool, lift it up to show how color drains out until it is cooked in.)
E. Rinse wool in clear water, again not shocking it by temperature changes.
F. Dry wool on screens or on something which allows the air to circulate.

Cornhusk Dolls

Cornhusk dolls are generally better for the free-choice activity time rather than the lecture rotations. The history of different styles of dolls and the use of cornhusks for other objects, such as sandals and baskets, can be fascinating.

Equipment:

1. Dry, brown cornhusks
2. String
3. Several colors of yarn
4. Scissors
5. Bucket of water

I. Preparation of Husks

A. Fairly dry, dead husks must be used.
B. Soak a large quantity of husks in a bucket of water to make them soft.
C. If desired, they can be bleached to remove stains. Husks also can be dyed various colors.
D. The silk is saved to make doll hair.

II. Procedure (See Figure 5.3.)

A. Choose and trim about six nice husks from the stalk.
B. Tie them together about an inch from the cut end (Step 1).
C. Carefully fold the husks over the string so the string and cut ends are hidden in the husks (Step 2.).
D. To form the head, tie another string about an inch down from the folded end (Step 3).
E. Make the arms by folding a cornhusk piece over a pipe cleaner or by braiding strips of husks. Slide the arms through the middle of the tied husks, right below the place where the string forms the neck of the doll (Step 4).
F. Then let your imagination run wild. A skirt with an apron can be formed and tied around the waist. Shawls, bonnets and hair can be added. Legs also can be formed and accessories added.

Soapmaking

Equipment and Materials:

1. Two pounds lard (not vegetable oil)
2. Eight tablespoons lye
3. Two cups cold water
4. Three half-gallon milk cartons, or some other appropriate mold

I. Procedure

A. Melt the lard and let it cool for about ten minutes.
B. Completely dissolve the lye in cold water, stirring about five minutes. Lye burns, so be careful not to splash it.
C. Stir in the melted and slightly cooled lard for about ten to fifteen minutes.
D. Pour into molds. (Half-gallon milk cartons that are cut lengthwise make excellent molds.)
E. Let the soap harden for twenty-four hours before cutting it into bars.
F. Let the bars dry out or *cure* about two weeks before using. The curing process actually allows the lard to kill the lye so it is no longer harmful to the skin.
G. For colored soap, melt a couple of wax crayons or candle color in the lard. After the lard is added to the lye, scents can be added to the solution. Any commercial, oil-based scent works, although candle scents work best. Ginger powder works well as a scent and adds a tan color.

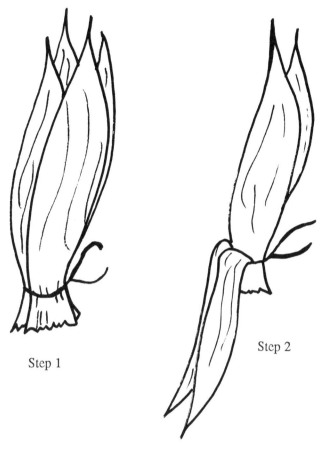

Step 1

Step 2

Figure 5.3

Step 3

Step 4

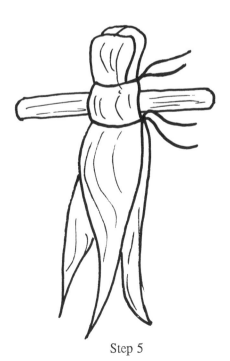

Step 5

Step 6

II. What two main ingredients were used by pioneers to make soap?

A. Lard, an animal fat, was obtained by killing animals, usually pigs or cows, and cutting away their fat. It also was obtained by collecting the drippings from the animal being cooked and straining out the impurities through cheesecloth.

 1. Lard from mutton and hogs was used for cooking. Beef lard and other refuse fat, which was not good to eat, was used in making soap.
 2. Beef fat, old bones, and undesirable pieces of meat were boiled in a soap kettle. As the grease came to the top, it was skimmed off with a brass skimmer or a wooden paddle.

B. Lye came from wood ashes which were collected from the fireplace and placed in a wooden trough having a screen bottom. The trough was placed outside and rainwater seeped through the ashes which leeched out the lye. The lye solution was collected in a crock, which was placed under the trough, and was said to be strong enough if an egg could float on it.

C. Soap was a luxury, not a necessity. Collecting the ingredients was very tedious so pioneers made large batches at a time and used it sparingly.

III. How many kinds of soap can you name that are in your house?

A. Hard soap, facial soap, hair shampoo, rug shampoo, dog shampoo, bathroom cleanser, laundry detergent, dish detergent, dishwasher detergent, etc.

B. How many kinds of soap were in a pioneer household? (One. Lye soap was used for every type of washing, including clothes and bodies. However, it could be found in two forms: hard and soft.

 1. Soft soap, which was stored in a barrel in a cool place, was used most often. Soap barrels were made of good, heavy staves since the lye in the soap would eat at the wood until it became soft and the barrel could no longer be used.
 2. Most people did not bother with making hard soap since it took a higher quality of lye and was more trouble to make than soft soap.

Native American Life Fair

Objectives:

1. To give students a chance to experience phases of typical daily activities in the lives of woodland Native Americans.
2. To help students discover some difficulties Native Americans faced while trying to deal with the arrival of white settlers to their land.

Structure:

The Native American Life Fair is a two-hour class consisting of a short introduction, five fifteen-minute demonstrations, and a council meeting in which students role-play tribe members who are considering a treaty offered by the U.S. government.

The class can involve one group of twenty students, taught by one instructor, or as many as three groups of sixty students, taught by three or four instructors. In the larger group, students are broken into four subgroups which rotate through the five demonstrations. Outdoor education staff members might teach the more technical demonstrations, while other leaders or teachers lead the gambling or war games activities. Demonstrations can include any five of the activities discussed in the following pages.

I. Introduction
Give students a short description of the Native Americans who lived in the area during the time of the earliest white settlements. Their history can be sketched in the form of pictographs that are drawn or burned on leather in a spiral pattern.

For example, the following story uses pictographs to tell the story of a small tribe of Potawatomi Native Americans who lived far in the north until two hundred years ago. They had a happy and healthy village with much meat on their drying racks ⟨symbol⟩ until one day a tribe of enemies began a war for their land ⟨symbol⟩. From the time the sun came up until it went down ⟨symbol⟩ the Potawatomi fought bravely until their chief was wounded ⟨symbol⟩. They fled as fast and as far as they could, moving their campsites each night ⟨symbol⟩. Then they had nothing and had to trade for all their food and clothing ⟨symbol⟩. They lived this way for a year ⟨symbol⟩ until the snow came ⟨symbol⟩. They were forced to set up their village but there was not food ⟨symbol⟩. The winter was hard. Many were hungry ⟨symbol⟩; many died ⟨symbol⟩. But spring came and they moved once again to a place with many trees ⟨symbol⟩ and a beautiful lake ⟨symbol⟩. There they established a village and once again they were healthy and happy with much meat ⟨symbol⟩. The hunting was good ⟨symbol⟩. There were bear ⟨symbol⟩, deer ⟨symbol⟩, birds ⟨symbol⟩, and fish in the lake ⟨symbol⟩. They were powerful ⟨symbol⟩ and strong like the buffalo ⟨symbol⟩, until one day their scouts saw ⟨symbol⟩ the "jemmo-ke-mon" or "long knives" ⟨symbol⟩, white men, ⟨symbol⟩. Their scouts saw wagons and women and children and plows. The hunting became poor once again ⟨symbol⟩. Their old people were hungry and many died ⟨symbol⟩. The long knives ⟨symbol⟩, they felt were very bad ⟨symbol⟩. They wanted the Potawatomi to make their marks ⟨symbol⟩ on pieces of paper that said they would have to give up their land and move west to a new place ⟨symbol⟩.

The pictographs should be read in a spiral pattern. (See Figure 5.4.) The story ends at the "present" year 1830. Afterward, students have a council meeting to decide what they should do about the white man moving into their area: sign the treaty and move west or refuse to sign and fight for their land?

II. Pictographs

A. Since there was no alphabet resembling our modern alphabet, a tribe's history usually was preserved in the form of pictures. One person in the tribe, known as the picture-writer, recorded one or several significant events of each year by adding symbols onto a piece of leather containing the tribe's history.

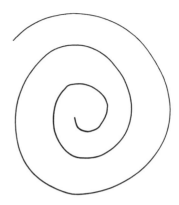

Figure 5.4

Figure 5.5

Drying racks	War	Sunrise/Sunset	Wounded chief	Camp sites
Food/Clothing	One year time	Snow	No food	Hungry
Died	Trees	Lake	Much meat	Good hunting
Bear	Deer	Bird	Fish	Powerful
Buffalo	Scouts	'Long knives'	White men	Poor hunting
Old people died	Wagons	Bad	Make marks	New place

B. Pictographs can be used as a demonstration or follow-up activity. Posters of various symbols obtained from books on the topic, or symbols created by students are made. (See Figure 5.5.) Let students use the symbols to make up a story, beginning in the center and spiraling outward. Students then share their stories with the group.

C. Students are given strips of cloth to design their own headbands using their Native American names on the pictograph symbols.

III. Sign Language

A. Sign language was a trade language between tribes of Native Americans as well as Native Americans and whites. Over a hundred tribes lived in the United States, and each had its own language. Communication between tribes desiring to trade, unite together, wage war, or make treaties often was necessary. Sign language was fairly universal, quite simple, and expressed only the necessary information. For example, there was no way to communicate words such as sparrow, hawk, robin, or eagle. A bird was either a big bird or a small bird.

B. Common signs are obtained from books (i.e., *Indian Sign Language,* Dover Press, by William Tompkins) or made up by students. Students choose their own Native American names, learning their signs. Names usually were given at birth and kept only until children could earn or choose their own. Young boys often would go on a three-day *vision quest* to fast, meditate, and wait for a sign or event which would give them their names. Native Americans chose their names to tell of courageous deeds they performed, animals they were like, or animals they wished to be like. Sometimes names were given by others who wished to comment on an event, such as the ''Man-Afraid-of-His-Horse.''

C. In this activity, students are shown and encouraged to choose one of the numerous common animal signs for their names. Next, they are shown the signs for adjectives which describe their animals: sleeping, big, little, running, striped, spotted, angry, wild, crazy, red, blue, etc. Signs of inanimate objects can be given as well: lake, flower, star, rain, river, etc. After students choose names, they pair off and exchange names by using the signs.

1. One asks another's name by waving a hand to indicate a question, pointing to the other person, and then moving the back of the hand away from the mouth in a downward motion. (See Figure 5.6.) The other person replies by pointing to himself and moving the back of the hand away from the mouth. (See Figure 5.7.)

2. After students choose names, they can begin to trade. The sign for trade is an X, which is made by crossing the forearms. Numbers up to ten are shown with the fingers. Tens are shown by peeling off the fingers of one clenched hand with the other. An animal skin is shown by pinching the skin on the back of the hand. Refusal of a trade is shown by bunching the fingers of one hand at the mouth and jerking the hand away as though spitting out bad words of the offer. Acceptance is

Figure 5.6

''How   are you   called?''

Figure 5.7

''I   am called...''

Figure 5.8

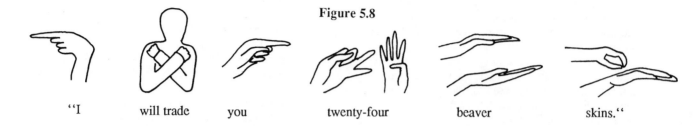

"I will trade you twenty-four beaver skins."

shown by extending the fist. (See Figures 5.8 and 5.9 for a sample trade.)
3. Students pair off and try trading with one another in sign language.

IV. Trapping

Students get actively involved with this demonstration. After a brief explanation of each trap, they practice setting the demonstration traps or try building their own.

Native Americans may have preferred traps to hunting for a number of practical reasons. By setting many traps, they did not have to spend time hunting animals, which may have taken hours. Traps also saved arrows which were often lost when a hunter missed game. Last, trapping did not damage the hide like the arrows of a hunter.

Traps were set along animal trails where tracks and droppings had been observed. Many traps had to be set and then checked daily in order to catch just a few animals; therefore, trapping still took much time and effort. A trapper never counted on catching one animal in each trap and may have set twenty or more traps for each animal obtained. It was unnecessary for the trapper to spend a great deal of time disguising traps to blend with the surroundings because animals did not notice its appearance. However, much attention was given to erasing the human odor from the area. A trapper may have spread a previous kill's blood or urine around the trap or set a few green sticks on fire, letting the smoke permeate the trapping area.

Trapping was originally a method of obtaining food and hides for Native Americans' personal use; they hunted and trapped only what was necessary. When white men came — offering knives, blankets, hatchets, guns, and other supplies in trade — trapping provided a source of income for Native Americans who trapped as many animals as possible. The Native Americans quickly learned to depend on what white men offered and were no longer as self-sufficient as they had been for centuries.

A. The Snare Trap — The snare trap is used for small animals such as squirrels and rabbits. A young sapling of sufficient size is located along a trail and stripped of its leaves to cut down air resistance when it springs up. A crossbar, which is located between two stakes in the ground, has a square whittled in the middle. This square holds the trigger, which is notched to catch under the crossbar. (See Figure 5.10.) The animal is caught in a sinew or rawhide noose, which is tied from the top of the sapling to the trigger and lies in a loop in front of the bait. The trap's success depends on the animal placing its head through the loop to get the bait and bumping the trigger loose, which causes the sapling to fly in to the air and hang the animal.

B. Deadfall Trap — Designed to kill a large animal such as a deer or bear, the deadfall trap should be made in miniature for students. A true deadfall trap consists of huge, heavy logs, which are pulled in place by a number of men or horses. More than once a Native American fell prey to his own, huge trap because it was difficult to set and easily triggered. It is made by precariously resting numerous logs on a crosslog, which is suspended between two trees. The bait, which might be part of a rabbit caught earlier in a snare trap, is suspended from a branch on the crosspole that falls when bumped. (See Figure 5.11.)

Figure 5.9

"You trade me five knives."

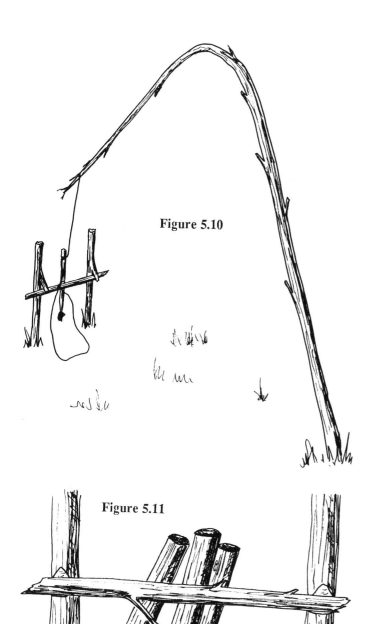

Figure 5.10

Figure 5.11

C. The Pit Trap — This trap also is made for large
animals and simply consists of a huge pit
camouflaged with small branches and leaves spread
across the top. An unwary animal steps and crashes
on the small branches and leaves into the pit below.
Sharpened stakes may be placed at the bottom of
the pit to impale the animal when it falls. A
miniature pit can be dug for students to practice
camouflaging. Perhaps they can even try to trap a
student from the next group.

V. Gambling Games

A. Sticks and Stones — This is a game of skill and
accuracy. Each of two players starts with five sticks
(more or less, as long as the number of players is
equal). The players sit several feet apart on the
ground and place one of their sticks in front of them.
They then take turns tossing a small stone back
and forth, trying to hit the other's sticks. A hit
counts only when the stone hits a stick on a fly. If
the stone rolls into or bounces onto the stick, it
does not count. When a stick is hit, the hitter takes
the stick and the opponent places another in its
place. The game continues until one player loses
all his sticks. Variations can be made by players
using larger sticks and stones and standing farther
apart. Distance can be increased or decreased, as
desired.

B. Deer Antlers — In this game of luck, ten disks
(cut from deer antlers, a stick, or ten small stones)
are colored on one side, and thirty pieces of corn,
pebbles, or any small object are placed in the
middle. Players then take turns tossing the "dice."
If five colored sides turn up, the player, who is
tossing, takes two pieces of corn from the middle
pile. If six turn up, the player takes four pieces of
corn. Seven colored sides earn six, eight earn eight,
nine earn ten, and ten earn twelve. (Score can
vary.) When the middle pile is gone, the game
ends, and the winner is the player with the most
kernels of corn. The game, however, can continue
if the winning players all pay the correct amount
from each of the other players winnings. If other
players are unable to pay the full amount, they are
out. The last one left or the one with the most corn
at the end of the game wins.

C. Corn Game — Two players place fifteen pieces of
corn in a pyramid shape (five on the bottom row,
then four, three, two, one), like bowling pins. Then
take turns removing as many pieces of corn from
any one row as they desire. Corn may not be taken
from two different rows during the same turn. The
object is to force the other player to pick up the
very last piece of corn.

D. Moccasin Game — This simple game of luck and
skill has many variations. One player hides a small
object under one of two moccasins (or similar
hiding spot), and other players try to guess which
moccasin hides the object. Two players can take
turns trying to outsmart the other. The player who
guesses correctly gets a piece of corn from a pile
in the middle. The player with the most corn at the
end wins.

E. Dice — This game of luck needs six flat stones or
chips of wood. One side of all six is painted black;
other sides are painted so there are three of one
color, two of another, and one of a third color.
Four colors are then possible. To place bets, players
simply state their bet and guess which color and
how many of that color shows up. A bet might be
for three black; therefore, the colors of the other
three do not matter. The best is won if that color
and proper number turns up. Any number of players
may place a bet on one roll; thus more than one
wins.

VI. War Games

A. Circle Stones — This is a game of judgment and skill. Any number of players stand at an equal distance with their backs facing toward the center of a large circle which is marked in the dirt. Players take turns tossing their rocks over their shoulders while staring straight ahead. The players who come closest to the center of the circle win. A variation is to have players face the circle and toss their rocks at the target area.

B. Flinch — In this game of reflexes, any number of players make a circle and stand with their arms folded. A person in the center tosses a medium-sized piece of wood or leather ball toward different players, varying directions and people. The tosser may fake a throw at any time. If players in the circle move their arms as though to catch the object and the throw is a fake, they are out. Even a flinch may eliminate players on a fake throw. If players drop the object or fail to move when the throw actually takes place, they are out.

C. Knee Coup — The bravest deed for Native Americans was to count coup. To do this, they galloped past the enemy at full speed, simply tagged them, and returned to their own forces without getting killed. Knee coup is another bravery game to prepare young warriors. The game lasts only a minute or two until a leader yells "time." The object is for each person to tag as many people on the knee as possible in a given time. For each knee they tag, one point is gained. The one who tags the most knees wins. For a more challenging version, count one point for each time a player tags a knee and subtract a point for each time that same player's knee gets tagged. Thus, a player who gets tagged five times and tags someone three times ends with a score of minus two. No boundaries are needed and any number can play. Keeping score is difficult, so short playing times are recommended.

D. Buffalo Robe — Like many Native American games, this game prepares young people for competitive interaction. A buffalo robe, such as a burlap bag or poncho, is spread out on the ground. Members of the group grip arms and circle the robe. It is best to grip wrists and hold tight because the object is to get others to step on the buffalo robe or break their grip. Anyone touching the robe is out. If there is a break in the circle, players on each side of the break, regardless of whose fault it is, are out. The game continues until three or four players remain around the robe. If one winner is desired, the battle can continue?

E. Spear Game — This game tests the strength of grips. Two players stand facing each other, each holding the same stick (four feet or so in length) with both their hands. The stick is held above their heads and one player's hands are on the inside of the other's. Players slowly lower the stick to waist level. Thus, someone's grip must slip to allow the stick to slightly rotate as it comes down. The player who allows the stick to slide, even slightly, is the loser.

F. Message Game — Usually each village had a fastest runner who was the village messenger. Messages were carried from one village to the next through runners and were passed on, pony express style, over hundreds of miles, reaching every village along the way. The game is played with any number of players, who are divided into tribes of three to six or eight or more. There should be two tribes or more. Tribes space their players long distances apart in horizontal lines. Perhaps they are even out of sight from some other players when the lines follow a trail. The tribal lines should be parallel with one person from each tribe standing near the other. The leader gives the first person in each line the same message, such as "Five hundred white men are three days' journey away." On signal, people run to those people next in their line and deliver the message. This continues down the line. The last person in each line delivers the message to the leader. The first tribe with the most accurate message wins.

G. Tomahawk — Two sticks, representing toma-hawks, are placed twenty to forty feet apart on the ground. Two warriors then are placed in the middle, each holding opposite ends of one stick. The object is for one combatant to drag the other to his own tomahawk. If a player loses his end of the stick, the other wins. Also, whichever player reaches his tomahawk wins. When playing for real, the loser might lose more than a game.

VII. Wigwam Construction

Woodland tribes of the midwest regions did not live in tipis like the nomadic Plains tribes. Plains tribes needed a form of shelter which was easy to construct, dismantle, and move to a new location as they followed the great buffalo herds. Tribes of the woodlands lived primarily in wigwams, wood-framed huts shaped much like inverted bowls. Although simple and quickly constructed, wigwams provided a more permanent form of shelter than the tipis and suited the semi-agricultural lifestyle of the woodland trips.

Lodges were made of materials found in the surrounding forests and marshes. Limber poles were used to make sturdy frames, shaped like loaves of bread and were tied together with strips of tough bark. Certain kinds of trees had bark that was peeled off in large sheets to cover the lodge frame. These sheets were lashed on the frame and overlapped so the lodge was watertight. Animal skins also covered the lodges.

The best materials for the frame are young brush willows which grow to a height of about sixteen feet in low, wet areas. They are straight and limber with only a few small branches. Although willow is best, any young, straight, limber tree works. A wigwam needs about forty poles, from twelve to sixteen feet long, which must be cut, trimmed of branches, and dragged to the construction site. It is best to use poles the day they are cut because they dry quickly, losing their flexibility; thus the bark used for tying does not come off easily in long strips. If necessary, poles can be stored in a pond or stream for about a week.

The bark can be peeled off with just the fingers. The longer the strips, the better because they are used to tie the lodge form together. Bark strips are piled in the shade or kept in a tub of water to keep pliable. Leather, twine, or string can be used as a substitute, but bark does not rot as fast and is more authentic.

The longest and straightest poles are placed on the ground in a circular or oval pattern. Use a stake and mallet to make holes in the sod for pole bases. The holes can be about two to three feet apart; however, the radius of the circle or oval must be small enough so the tops tie together, forming a wicket.

Decide where the door can be, or have a door at each end. Bend the upright pole beside the door to its opposite pole at the far end of the lodge. Tie them together at the decided roof height. Repeat this with the pole on the other side of the door so there are two large, parallel wickets. Proceed by tying opposite poles across the top of this frame, at right angles to the first two wickets. While lashing poles together, it is important that the ceiling is the desired height and the limber poles are conforming to the desired shape. When all the vertical poles are lashed together, horizontal poles may be tied in place, starting with a bottom row about two feet above the ground. In some cases, horizontal poles can be woven into the verticals, giving the frame better shape and stability. Leave an opening for the door. As all of the frame's intersections are secured, the frame begins to feel sturdy.

Native Americans used large sheets of birch bark for lodge covering. Sometimes they combined birch with elm bark or woven rush mats. Now that elm and large birch trees are rare, mats of cattail leaves or even cornstalks are substituted. In sewing cord through cattails, be sure to considerably overlap them because they shrink while drying. A square hole, attached to two parallel sticks with cords to cover it when raining, is left at the top center so smoke can escape. This same type of framework, covered with a canvas or nylon tarp, can be used as a sweatlodge.

VIII. The Sweatlodge

The sweatlodge had great spiritual and medicinal significance. A *sweat* usually was administered by a holy man of the tribe, a priest, after prayers and meditation by him and a participant. The sweat often was used to cur illnesses because the cleansing process freed the body from impurities. It also was used to prepare young boys for the vision quest in which they sought the path to manhood and a purpose for their lives. The sweat enabled them to free themselves of evil and to dedicate themselves to the spirit. Often a brave requested a holy man to give him counsel in the sweatlodge before a significant battle or an important decision. During this time, both meditated and sought counsel regarding the good of the people.

The frame of the sweatlodge is built like that of the wigwam. The covering, however, needs to be a large sheet of plastic or a tarp, preferably black. The Native Americans used animal skins. A pit to hold rocks should be dug in the middle of the lodge floor.

To prepare for the sweat, a fire, about twenty feet from the lodge, should be built well ahead of time. The fire can be layered with wood and rocks, built over rocks, or rocks can be placed in the fire after it is burning. To obtain the most heat, rocks should be in a good, strong fire for over an hour. Rocks about the size of a softball are best, since they are manageable and retain heat well.

When all is ready, participants silent and dressed in bathing suits or shorts and t-shirts, enter and sit in a circle around the inside edge of the lodge. The tarp is pulled down around the edges so there is no draft. One person with a shovel or pitchfork then takes rocks from the fire and slides them under the tarp into the pit. Care is taken not to include embers from the fire or participants are smoked out of the lodge. Inside, a person splashes water from a large bucket onto the hot rocks, as desired, which creates the steam and a great sauna effect.

For further excitement, people emerging from the sweatlodge can throw cold water on each other or, in winter, take a roll in the snow.

This may not be a good class for typical student groups, but is a great alternate activity for special groups.

IX. Native American Cooking

The Native American cooking class is taught best around a fire, outside or in a classroom fireplace. Students participate by shelling or grinding corn, stirring parched corn, or preparing plants for tea.

A. If desired, historical information can be covered while cooking.

1. Foods of the plains and woodland varied slightly. Plains tribes depended heavily on meat from wild game, only supplementing when they could by gathering greens, roots, and bulbs in season. The woodland tribes also ate much meat, but were more agricultural, remaining in one place from planting until harvesting.
2. Traders and settlers added to the Native American diet by bringing coffee, white sugar, flour, and new vegetables. As traders came, cooking slightly changed as well since, through trade, Native Americans had metal knives, iron pots, and skillets.
3. Native American mealtime concerns were twofold: What would be eaten today? What would be eaten this winter? Therefore, food preparation included both concerns, cooking for the present and preserving for the future.

4. Native American women were responsible for all the food preparation from planting to harvesting to cooking, from skinning to frying, from woodgathering to firebuilding. Men hunted, which was the extent of their contribution. Among woodland tribes, women were the first North American farmers.

5. Native Americans used at least three methods for preserving foods.

   a. Drying strips of meat, vegetables, and fruits over a rack in the sun was a slow process, but insured food would be available at a later date. The cooks had to be constantly aware of the drying racks, so the strips of food were not stolen by ranging camp dogs or by young boys counting coups in their war games.
   b. Salting also worked for meats and fish, but could be done only in areas where salt was plentiful.
   c. Smoking was slightly different than drying because it utilized a smoky, greenwood fire and added a unique flavor to the meat.

6. Woodland tribes depended heavily upon garden crops such as pumpkin, squash, zucchini, and corn. Corn was an ideal crop since its cultivation did not require vast open spaces and a long, hot growing season like other grains. Even now most other countries do not grow corn for human use, a fact which illustrates the uniqueness of Native Americans' contribution to early settlers.

7. Fewer ingredients, spices, and seasonings were available.

   a. Milk was unknown to woodland tribes, although tribes of the southwest may have utilized goat milk.
   b. The only sweetenings were the natural flavors of berries and fruits, maple sap boiled into sugar or syrup, and honey.
   c. Eggs were not commonly used, although various bird eggs were available, particularly those of the large wild turkey.
   d. Salt was available to woodland tribes. Found along rivers and streams, it sometimes required boiling down from a water solution. Plains tribes had much more difficulty obtaining salt.
   e. Some herbs, wild vegetables, and greens were gathered and added to meat and soup dishes for extra flavor.

B. Some items are cooked during class with the students' help. Others are discussed and described while cooking.

   1. Cornbread is baked in a Dutch oven. A frybread is made by following a cornbread recipe and frying the batter like pancakes or in deep fat. Students grind some field corn

between rocks and even add the results to the cornmeal batter. Sometimes a flour, made of ground acorns or seed pods from cattails, were added to the cornmeal to extend it.

   2. Sumac, sassafras, white pine needle, or yarrow tea can brew on the fire. Sweeteners of brown sugar, honey, or maple syrup can be added.
   3. Dried apples or other fruit can be suspended over the fire from a crosspole. For best results, prepare the apples ahead of time. Core and slice apples so sections resemble doughnuts, then dry them on a cookie sheet for three or four hours in an oven set at 150 degrees. They then are strung on the crosspole for class. Meat is done much the same way, cut in thin slices and dried for a longer period of time.
   4. Pemmican was a common scout food that was taken on journeys. Dried meats were ground into a powder and melted animal fats, and dried, ground fruits were added. The mixture then was stuffed into the linings of animals' intestines much like sausage. It was easily carried, non-perishable, and high in nutrition.
   5. Parched corn is made from field corn or dried sweet corn. A very small amount of fat and a cup or so of corn kernels are put in a cast iron skillet. Stir constantly over a hot fire, until the kernels are toasted brown, dried, and crunchy. It is a good treat, but a bit tough on the teeth.
   6. Stone soup is a challenging yet authentic dish which was prepared by Native Americans in the stomach lining of a large animals or in bowls made of wood. Any kind of soup mixture is used, including raw beef chunks, with any type of raw vegetables. Some seasonings or broth also can be added. Heated stones are removed one by one from the fire and dropped into the soup. As the stones cool, they are fished out of the soup and returned to the fire for reheating. Within approximately twenty minutes, the soup boils and, with continued simmering, is completely done in less than half an hour. The results are amazing and impressive. This cooking method was used before the white traders' iron pots and kettles were available, since Native Americans did not have containers that could sit directly on fires.

X. Some Tipi Odds 'n Ends

A. The tipi was a portable home which Native Americans considered a sacred place. The dirt floor symbolized the earth on which the Native Americans lived. The sides of the tipi, vaulting to a peak, symbolized the sky, and the roundness was a reminder of the sacred circle which had no beginning and end.

B. The tipi, both very sturdy and light, was erected in minutes.

C. Women made tipis out of poles and skins. A tipi could be made with as few as ten, or as many as thirty poles. The covering was made from any number of pelts, depending on the size of the tipi and the pelts.

D. The larger the tipi, the greater the wealth of those who lived within. A bigger tipi needed more horses to transport it. Horses were a sign of wealth.

E. Tipis were not only erected and taken down by women, they also were owned by them.

F. Tipis always were erected with their entrances facing east because of prevailing winds and the morning sun's warmth.

G. If the flaps on the door were closed when approaching the tipis, one had to make his presence known before entering by saying something or rattling beads hung near the door. If the flaps were opened one was welcome to enter.

H. Upon entering a tipi, women entered to the left and men to the right. Thus, all sat down. Women were on the right side of the host, men on the left. The host sat directly opposite the entrance.

I. Women kept their belongings on the left side of the tipi from the opening. Men kept theirs on the right.

J. One never passed between a seated person and the fire. It was considered impolite.

K. Smoke flaps at the top of the tipi were adjusted to accommodate weather conditions. In the warm summertime, smoke flaps were opened wide and sides of the tipi were rolled up to cool it. For additional warmth, more layers of skins were hung around the inside.

L. Insides of tipis often were decorated by men who painted their dreams on the walls. Some paints they used were:

   1. blue from dry duck manure and water,
   2. yellow from bullberries or buffalo gallstones and water,
   3. black from burnt wood,
   4. green from plants,
   5. white from certain clays.

M. Fire in the tipi was used for cooking during inclement weather. Much of the cooking was done outside, since an inside fire is almost always a bit smoky.

N. The grease and smoke from the tipi fire rose and stuck to the hides covering the tipi. When they became old and weathered, new coverings were made and the old coverings were used for making moccasins. Grease and smoke made the skins tougher and more water-resistant.

XI. Council Meeting

Students gather in one large group and get various roles to play. As they are introduced, students are painted with a water-base war paint. Students taking a role may be given a name or may use the name they chose in the sign language activity. Some may be named after actual prominent Native Americans of the tribe. Roles may include:

1. A scout who has traveled extensively, even to the west side of the Mississippi, and has seen that the land offered by the treaty is good
2. A chief who is wise and well respected by the tribe (The chief receives a necklace, peace pipe, and *talking stick.*)
3. A hunter who knows the land is becoming bare of animals and is concerned about moving his people to better land
4. A medicine man of visions and prophecies who is consulted by everyone
5. An interpreter who can read, write, and speak English
6. Tribal members who are called upon to discuss and decide whether or not to sign a treaty

To begin the council meeting, the instructor hands the talking stick to the chief, explaining its use. The talking stick gives a person the right to speak about the issues at hand. The chief is the only one who can hand the talking stick to another person. If someone wishes to speak, he simply, yet quietly, stretches out his hand. The chief then chooses whether or not to acknowledge him. Speaking without the stick brings the offender great disgrace.

At this point, an instructor attired and behaving as a U.S. government agent, enters and reads a treaty that the government is offering this particular tribe of Native Americans. The treaty is as authentic as research allows regarding actual dates, places, and people of the area. It may offer the Native American tribe some far-away land, blankets, weapons, food, and anything else appropriate.

The chief then is instructed to discuss the offer with his tribe and decide whether they sign the treaty and move, or stay in their own village, not signing the treaty and taking changes on the final outcome. The discussion may be quite lengthy and vehement; and, at any time, the government agent may press for a decision.

The council ends with the chief's report of the decision. The instructor concludes by sharing actual historical facts concerning similar treaties and the difficulty involved in making decisions. It should be stressed that there is no right or wrong answer regarding the treaty signing. The purpose of the council meeting is to let students realize the complexity involved in any treaty decision.

Pioneer Tools

Objectives:

1. To help students analyze the use of early homesteaders' tools.
2. To discuss with students the history of the Midwest frontier.
3. To give students a chance to use old tools and make a few items which pioneers may have made with them.

Equipment:

A number of authentic, antique tools are necessary for this class to be effective. These can be bought at antique stores, flea markets, and other such places. The wider the variety, the more the class can do.

Preferred:

1. Felling axe, single- and/or double-bladed
2. Broad axe and/or broad hatchet
3. Auger
4. Froe
5. Maul or mallet (This can be made in class.)
6. Stock of logs with various lengths and diameters, preferably oak, cedar, cherry, hickory, or other types which split easily, not elm or box elder
7. Several saws, old or new, but useable

Optional:

8. Mortise axe
9. Chisels, corner mortise chisel, mortise chisel, firming chisel
10. Adze, long-handled or hand adze
11. Draw knife
12. Shaving horse (This can be built with the pioneer tools.)
13. Sledge hammer and several wedges (These are not actually part of the class, but are useful in preparing wood for class. Pioneers would have used wooden wedges and a maul.)

I. The Class

This class deals with settlers of the wooded areas west of the Appalachians, who homesteaded in the late 1700s and early 1800s. The first half of class is discussion and demonstration of tools. In the second part, students divide into groups of three or four and rotate through various stations, examining and working with tools. One adult supervisor should be in charge of each station.

A. A felling axe is used by students, one at a time, under careful supervision. They try notching a log lying on the ground, if it is secured so it cannot roll. A ten-foot tree also can be buried in the ground for students to chop down.
B. A broad axe is used by students at the same station or at a different one. Broad axes are heavy and awkward, so students may have trouble safely using one. Thus, this may be only an observation station.
C. Augers are used to drill holes in a split log for bench legs.
D. The adze is used on a squared log at another station.
E. The froe and maul are used by students to split shingles. It is usually best if an adult holds the froe and a student uses the maul. Twelve-inch logs should be split in half prior to this activity.
F. Draw knives are set up with the shaving horse for students to use.

II. Introduction

A. When settlers arrived in the immense forests west

of the Appalachians, they faced a tremendous task. To homestead the claimed land, they had to remove huge trees which were some forty feet around, pull the stump from the ground, build a cabin and barn, put up fencing, and still take time to plant crops. All this was done with just a few basic tools.
B. Each tool had a specific purpose and each homesteader was a woodcraftsman, who was very particular about the care and use of his tools. A froe was not used to chop down a tree, nor was an axhead used as a hammer.

1. A rule of wood on metal or metal on wood is important. Metal should never be used on metal except in blacksmithing. It is dangerous and can ruin tools.
2. Edges should always be protected and never placed in the dirt or dug into the ground. A block of wood should be used at all times under a tool which may hit the ground while in use.
3. No students should be allowed to use a tool, except with an adult's supervision.

III. Building a Barn

Barns were usually made of oak since it was plentiful, strong, and easy to use. Roofing rafters often were made of tamarac since it does not rot as quickly as other woods. Shingles were usually cedar for the same reason, and cedar split easily. Barns were nearly as important to homesteaders as cabins since they needed a place to store crops, house animals, and store tools, implements, and seeds.

A. Axes were used to chop trees down to clear land, and oxen or horses dragged the logs out and pulled up stumps. Often settlers girdled trees on extra acres of land. They chopped a ring several inches wide around a tree so it would die. No leaves came out the next spring and enough sunlight got through so crops could be planted among the trees until settlers had time to remove them. The axe was the most important tool for woodsmen because it was totally indispensable. With an axe and enough time, they could build an entire homestead.

1. Axes had knife edges; they were sharpened on both sides for a V-shaped edge. Knife edges were on tools used for cutting.
2. Single- and double-bladed axes were used. The double-bladed axe enabled woodsmen to chop twice as long without returning to the grindstone. (See Figure 5.12.)

B. Broad axes were used to turn round logs into square beams. (See Figure 5.13.) Haste in building cabins usually dictated that logs simply be notched and stacked one upon the other with chinking of mud packed between. As time allowed, logs were squared and fitted together to make a frame skeleton for cabins or barns. This provided greater structural strength and a more permanent joint.

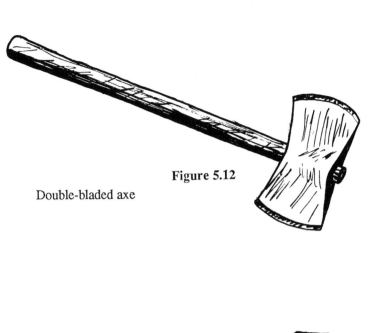

Figure 5.12

Double-bladed axe

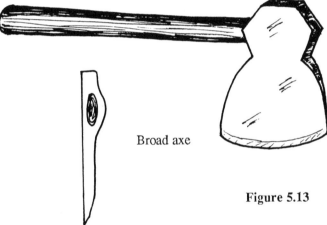

Broad axe

Figure 5.13

1. Four lines, marked on a log with a chalkline, indicated the four corners of the beam. (See Figure 5.14.)
2. A felling axe was used to *score to the line*. Woodsmen stood behind logs and chopped marks every few inches into them. These marks were cut into logs deep enough to meet the desired flat side of beams. (See Figure 5.15.)
3. The broad axe then was used to chop vertically through wood, chipping off sections marked by the felling axe. (See Figure 5.16.)

   a. The broad axe had a short handle and a very broad blade, much wider and heavier than the felling axe blade, so it could not fell trees. Most broad axes had slanted handles; therefore, the user's knuckles were away from the log and did not scrape as the axe was brought down.
   b. The blade had a *chisel* edge, an edge sharpened only on one side and used for shaping rather than cutting. It was actually more of a chisel than an axe and was designed for making a smooth surface. The flat side was placed against logs with the sharpened edge out to chip away sections of wood.
   c. One usually did not chop wood, but simply guided the axe downward, letting the heavy weight do the work.

4. As each side was finished, the log was rolled over to flatten another side until a square beam resulted. Hand-hew beams are identified by chop marks left by the felling axe as it scored to the line.

C. Augers and chisels were used to form joints which held beams together. Beams were not notched like logs in a log cabin. The end of one beam was fitted into the side of another by means of a mortise and tenon. The mortise was a square hole in the side of a beam and the tenon was a smaller square protruding on the beam's end. These made a tight fit which would have been impossible with round logs.

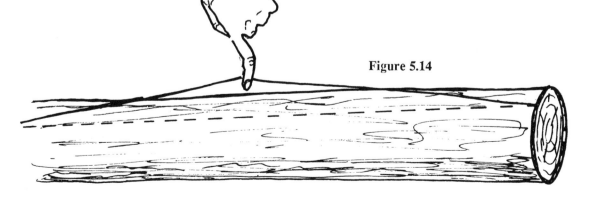

Figure 5.14

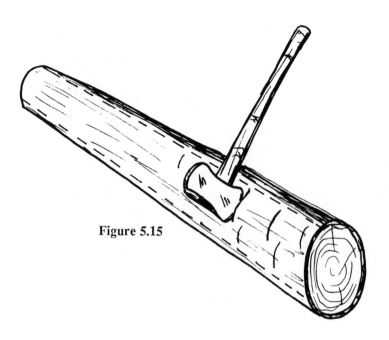

Figure 5.15

5. After placing the tenon end of the beam into the beam with the mortise, the settler drilled two holes through the joint which passed through the mortise and tenon. Wooden pegs were driven into these holes to pin the beams together. These were called *trunnels,* slang for tree nails. Wooden pegs were best because they did not rust and expanded and contracted like the beams, always insuring a tight fit. (See Figure 5.21.)

D. Horses and ropes were used to pull the pinned beams into place; and while they were precariously held, one brave barn builder climbed to the top and pegged the corner beams in place. With a froe and maul, the farmer then split plants for the barn's siding. Sometimes, a long pit saw was used. It required two men: one on top of the log, holding one end of the saw, and one holding the opposite end, standing in a pit below. Thus, they sawed the log into planks.

Figure 5.16

1. First an auger was used to drill two or three adjacent holes about halfway into the beam. This began forming the mortise. Augers had screw-type tips which bit into wood. They were difficult to sharpen and required a great deal of muscle power to use. (See Figure 5.17.)
2. A mortise axe, a type of chisel, was used to turn round auger holes into a square mortise. It was placed on the edge of the holes and hit on the head with a maul (wooden hammer). (See Figure 5.18.) By chiseling away at the sides of the holes, a square hole resulted.
3. Several other chisels, such as a wide-bladed firming chisel or a mortise chisel with a narrow chisel edge, were used to cleanup the mortise. A corner mortise chisel, with a blade in the shape of a right angle, was also helpful in making good, clean corners in the mortise. (See Figure 5.19.)
4. With a saw a belt was cut a few inches into a second beam near the end of all four sides. Using a froe or a chisel and maul, the settler chipped away the outer edges of the beam's end, leaving a square peg, six or eight inches in length. This was the tenon which was shaped to fit exactly into the mortise. (See Figure 5.20.)

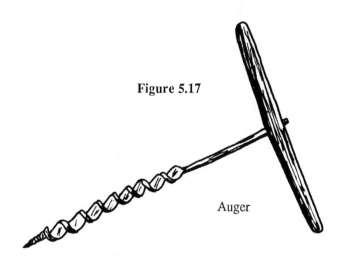

Figure 5.17

Auger

E. A froe was used to split, *rive,* shingles for the roof.

1. A froe was used like a wedge to split thin planks off small logs because it had a knife edge. First, a log about a foot long was quartered lengthwise, *billeted,* with a wedge. The froe then was placed on the log at an angle

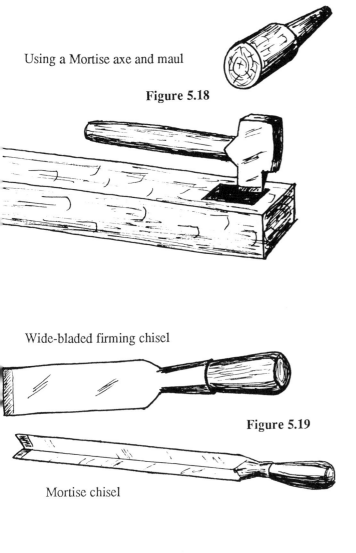

Using a Mortise axe and maul

Figure 5.18

Wide-bladed firming chisel

Figure 5.19

Mortise chisel

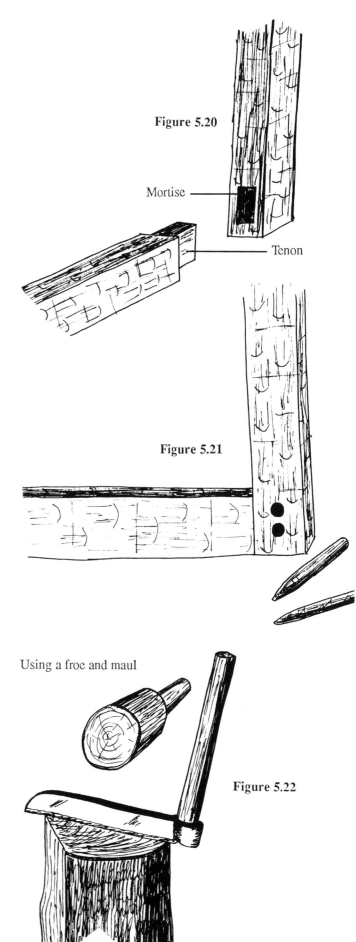

Figure 5.20

Mortise

Tenon

Figure 5.21

Using a froe and maul

Figure 5.22

from the corner to the outside edge, about
one-half inch in.

2. Using a maul, the froe was hit squarely on top
   in the blade's center. After several blows, a
   twisting motion on the froe handle snapped
   off a shingle. (See Figure 5.22.)

F. The adze and draw knife were used for smoothing
   beams, trunnels, flooring, doorjambs, tool handles,
   and furniture. Not much smoothing was done in
   the barn; these tools often were used for finer work
   in the house. Thus, they often were referred to as
   parlor tools.

   1. The adze was used for smoothing beams and
      flooring. It had a chisel edge and worked like
      a plane. The builder stood on the beam with
      one foot raised on the heel, toes up. He then
      brought the adze down and toward him,
      scooping into the wood and ending on the
      bottom of his foot. This stripped shavings off
      the wood, leaving a smooth surface. This tool
      is often called a toe or foot adze. (See Figure
      5.23.)

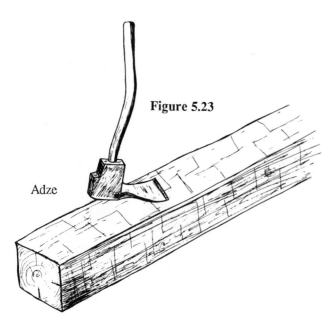

Figure 5.23

Adze

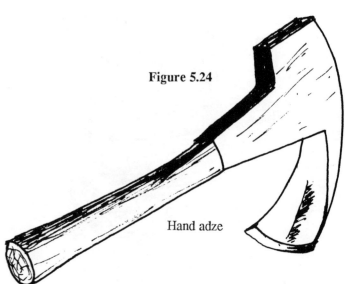

Figure 5.24

Hand adze

2. A hand adze was used in the same manner as the foot adze, but for smaller projects. It was important that more of a scooping motion was used rather than a chopping action. (See Figure 5.24.)

3. Planes with chisel blades also were used for smoothing. The old planes were very similar to those of today, except they were made entirely of wood, except for the blade. They came in all shapes and sizes, some being used for barn beams and some for fine furniture.

4. The draw knife was a type of plane. The craftsman put his hands on both handles and drew it along the wood toward himself. The blade was sharpened on only one side like a chisel, and the sharpened side was held with the flat side against the wood. (See Figure 5.25.) Somehow, the wood had to be braced so the draw knife could be pulled toward the user. Consequently, a shaving horse consisting of a bench and a foot-operated vice often was used. (See Figure 5.26.) The draw knife was used for fine work such as tool handles and furniture legs.

IV. Projects

Some projects can be done during the activity part of class or as a future activity. Good wood is necessary.

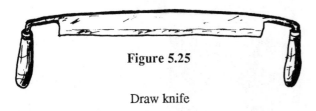

Figure 5.25

Draw knife

Seasoned wood works best since it splits easily and is not rotted. Ash, hickory, oak, cherry, and walnut, sassafras, elm, aspen, maple, and box elder are good because they do not split easily.

A. Mallet

1. Cut a six- to eight-inch diameter log about six to ten inches long. Drill a hole on the side; the deeper the better.

2. Cut a handle with a diameter a bit larger than the hole. Using a hatchet, taper the handle to fit. Next, pound the handle into the hole and put small wedges in it to hold the handle in place.

3. The hole can go completely through the head. Therefore, it is placed all the way through, the end split, and a wedge placed in the split end to widen it so the handle cannot slide back out. (See Figure 5.27.)

B. Maul

1. Cut a log of medium diameter about a foot or more long. With a saw cut a one- or two-inch deep belt all the away around and somewhere in the log's middle.

2. Split away pieces at one end around the outside with a froe and maul. This leaves a handle and makes a one-piece maul. (See Figure 5.28.)

C. Milk Stool

1. Split off a slab of wood from a foot-long log with a froe and maul.

2. Drill a hole almost all the way through the slab in the center of the rounded side.

3. Taper the end of a leg with a hatchet and wedge it into the hole, forming a one-legged milking stool. (See Figure 5.29.)

Shaving horse

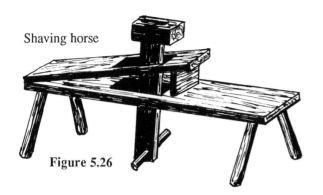

Figure 5.26

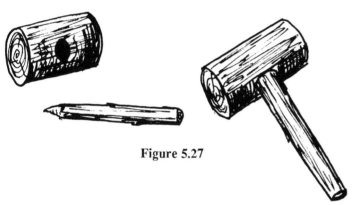

Figure 5.27

Figure 5.28

V. Alternate Class: Barnology

A barnology class, consisting of a scavenger hunt, can be offered if an old barn is on or near the site. Each student has a sheet of paper listing things to find such as a mortise, hand-hewn beam, square nail, trunnel, etc. These things are labeled with letters on cards in the barn. Students match the cards with their lists, writing the letters on a blank beside each item on the list.

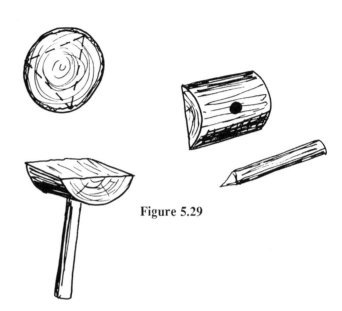

Figure 5.29

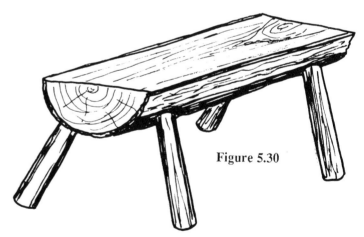

Figure 5.30

D. Bench

1. Split a log in half with a sledge hammer and wedges.
2. Drill four holes in the rounded side near each of the four corners. Holes should be drilled at an angle to make a wide base of support. Deep holes provide more strength.
3. Taper the ends of four legs and pound them into place.
4. Even the four legs with a saw so the bench is level. (See Figure 5.30.)
5. Three-legged stools are easier since only three holes need drilling and three points form a plane. Thus, the bench is easily leveled.

Michigan Country

Objectives:

1. To encourage students to work together in small family groups toward the goal of survival in a new land.
2. To help students experience the frustrations, challenges, and difficulties of homesteaders in the Midwest.
3. To give students a chance to learn surveying and mapping techniques as well as the skills of buying, trading, and selling.
4. To give students a feel for environmental issues as natural resources in the new land diminish throughout the year.

Equipment:

1. Printed money, one, five and ten-dollar bills
2. Surveying equipment: compass and twenty-five foot rope (one set per family)
3. Fifty to one hundred pelts (pieces of cloth, cardboard, or carpet squares) hidden throughout the land area of the settlement in bushes, trees, under brush, by water
4. Three white squares of poster board representing salt licks placed throughout the area, primarily near the water source
5. Cards representing the general store's stock:

   Plows, $5, ten cards
   Pigs, $2, ten cards
   Oak desk, $2, five cards
   Salt for one year, $2, ten cards
   Wagon, $2, ten cards
   Land, $4/acre, eighty cards
   Deer, $5, five cards
   Beavers, $1, five cards
   Tools/axes, $4, ten cards
   Sheep, $3, ten cards
   Lumber (for a cabin), $4, ten cards
   Flour for one year, $4, ten cards

   Gun and ammunition, $5, ten cards
   Barrel of water, $5, ten cards
   Corn seed for one acre, $1, thirty cards
   Firewater, $3, ten cards
   Nails, $2, three cards
   Horses, $3, ten cards
   Surveying equipment, $5, ten cards
   Trinkets, $2, ten cards
   Two oxen, $5, ten cards

6. Chance cards, which are drawn at the town meeting:

   Examples:

   *Land value will increase $1 per acre.
   *A flood wipes out all crops of anyone with land near water.
   *Land prices increase to $7 per acre.
   *A tax of $20 per family is levied to pay the mayor's salary.
   Other situations can be created.

7. A hundred or so deeds of land claims (See Figure 5.31.)
8. Two tally sheets per family (See Figures 5.32 and 5.33.)

The Michigan Territory

whereas _____ has requested to take up ___ acres of land at township marks _____ in the county of _____ for which he agrees to pay ___ per acre, paid in gold, silver or paper money of the United States of America. I further attest to the fact that the land has been properly surveyed, with the following landmarks: _____
In witness thereof: _____
                                                    surveyor general
To: _____
      Governor of Michigan Territory   *Date*

Figure 5.31

Michigan Country Tally Sheet
Figure 5.32

The First Year 1820

ASSETS

    1. Cash $ _____

    2. Equipment/Supplies:

        _____ $ _____

        _____ $ _____

        _____ $ _____

        _____ $ _____

                               SUBTOTAL $ _____ $ _____

    3. Land Value:

        Number of acres _____ times the amount per acre $ _____ = $ _____

    4. Crop Return:

        Number of bushels per acre _____ times price per bushel $ _____ = $ _____

                             TOTAL ASSETS $ _____

TAXES AND FINES

    A $5 fine is assessed for each of the following items that your family did not obtain during the first year (1820) and are subtracted from the total amount of assets.

    1. Food (includes one meat or sheep card OR pelt,
       plus one salt card or salt lick) $ _____

    2. Water $ _____

    3. Shelter (includes one acre of land PLUS written
       proof of one tree or a cabin card) $ _____

    4. Clothing (includes one cloth card or one pelt) $ _____

    5. Tax $ _____

           SUBTRACT TOTAL TAXES AND FINES FROM TOTAL ASSETS $ _____

                           GRAND TOTAL $ _____

This class works best with a fairly large number of students, possibly sixty to a hundred. It is designed for junior high students or older. An adequate number of adult leaders are needed for appropriate supervision and assistance; at least six adults are recommended. This particular class concerns the Michigan Territory of the 1800s, but particulars may be adapted for states where students live.

I. Introductory Discussion

    A. Michigan was settled in the early 1800s. This class is set in the years 1820 and 1821.

    1. Michigan was a territory purchased as part of the Northwest Ordinance in 1798.
    2. What was the land like? Were there Native American tribes? What were they like?
    3. Who were the pioneers and settlers? Why did they come? What did they bring with them? (See Homesteading class.)

    B. A frontier family arriving in the wilderness had some immediate survival needs:

        1. Water — Usually the family tried to gain a

piece of land that had or was near a water source.

2. Food — They obtained meat from hunting and supplemented it with plants they gathered until their crops were harvested. They also may have had some staples that they brought with them or, perhaps, they traded with other settlers who arrived earlier and had a year harvest. Salt was an important item, not just for flavoring, but for preserving meat. Anyone with a salt lick on his land had an important commodity. The salt licks found in class are very valuable and may be purchased by the general store owner.

3. Shelter — First, a family might live in a wagon or build a primitive lean-to until a simple one-room cabin was built. They added on to the cabin over the years.

4. Clothing — Minimal clothing was brought and more was obtained by using animal skins and furs, spinning wool, knitting, and weaving garments. However, sheep and wool may have been difficult to obtain at first. In this class pelts (pieces of cloth) are found throughout

Michigan Country Tally Sheet

The First Year 1821

Figure 5.33

ASSETS

1. Cash $ _____

2. Equipment/Supplies:

   _____ $ _____

   _____ $ _____

   _____ $ _____

   _____ $ _____

                                   SUBTOTAL $ _____ $ _____

3. Land Value:

   Number of acres _____ times the amount per acre $ _____ = $ _____

4. Crop Return:

   Number of bushels per acre _____ times price per bushel $ _____ = $ _____

                                                                   TOTAL ASSETS $ _____

TAXES AND FINES

A $5 fine is assessed for each of the following items that your family did not obtain during the first year (1820) and are subtracted from the total amount of assets.

1. Food (includes one meat or sheep card OR pelt, plus one salt card or salt lick) $ _____

2. Water $ _____

3. Shelter (includes one acre of land PLUS written proof of one tree or a cabin card) $ _____

4. Clothing (includes one cloth card or one pelt) $ _____

5. Tax $ _____

                         SUBTRACT TOTAL TAXES AND FINES FROM TOTAL ASSETS $ _____

                                                         GRAND TOTAL $ _____

the woods and are worth a great deal at the general store. The pelts may represent clothing or an item of food.

II. Roles

A. Students role-play members of a pioneer family.

1. Each family has $50 cash which was brought with them.
2. All families came from the Boston area.
3. Due to many problems encountered on the way, the family's only possession upon arrival is clothing.
4. Each family has four members. (The instructor divides the large group into groups of four.)
5. The family's goal is to simply survive for two years, gain at least one acre of land, and turn a profit by the end of the second year.

B. Adult leaders' roles are vital and should be played as realistically as possible, including costuming and acting.

1. Land Officer (Mayor) — The main coordinator of the class, the land officer, sells deeds to surveyed land and keeps track of them on a map of the area. He also runs town meetings, keeps law and order, and settles disputes.
2. Surveyor — This person runs a surveying class for families who wish to survey their own land. The class is offered for a minimal fee at the beginning of the first year. Or, families simply pay the surveyor a larger fee and he surveys the land for them. The surveyor is a rough, frontier type who is out to make money.
3. General Store Manager — As operator of the store, he sells typical general store stock, including all items represented by the cards. He sets the prices which are arbitrary and may be dickered over. He sells whiskey and trinkets and buys pelts, salt licks, and possibly other things including trash, litter, plastic bags, and whatever anyone can convince him to buy. He is basically a helpful, friendly person who enjoys dealing with settlers and wants to make money.
4. Highwaymen — Three or four highwaymen wander around the settlement, offering their services. They are docile and enjoy the settlers' goods, especially their whiskey. They know the land well — where pelts, salt licks, and water sources are found and are willing to trade this information for any of the settlers' goods. Students are not informed that highwaymen may lie, cheat, sell false land deeds, give misinformation concerning the location of pelts and salt, and cheat settlers out of as many goods and money as they can while giving as little as possible in return.

III. The First Year

The first year lasts twenty (or thirty) minutes. At the end of that time a town meeting is called by the land officer. Families must have proof of survival.

A. Each family must have a surveyed land claim for at least one acre. The family can survey it or hire the surveyor.

1. A stake is driven prior to class to mark the territory's center. Other stakes are placed in four cardinal directions, twenty-five feet out. These are used as reference points so families can survey land. All claims consist of a twenty-five-foot square acre, following the lines of the cardinal directions.
2. Families must begin surveying from one of the stakes or from the corners of others' land. Surveying is done with the use of a compass and a twenty-five-foot rope which must be purchased at the general store.
3. Corner markers, consisting of rocks, sticks, trees, etc., must be placed to indicate the claim once the land is surveyed.
4. At some point, the family must go to the land office to purchase the land and obtain a deed. Land deeds may be offered by highwaymen. However, unknown to students, they are false and cannot be honored at the end of the year.

B. Each family must have proof of a water source; they must have a glass of water to show the land officer. Water is obtained by purchasing land along a body of water, purchasing rights from another family or from highwaymen (who may not have the right to sell it), digging a well, or using other creative thinking.

C. Each family must have proof of clothing, represented by either a belt, which was found or purchased, or a clothing or sheep card, which has been purchased at the general store.

D. Families must have proof of shelter which is obtained by purchasing an acre of wooded land, a tree from another family with a wooded acre, a wagon, or the lumber for a cabin.

E. All families must have some food, a food card that has been purchased or a pelt representing trapped animals that have been used for meat.

F. Families should have enough assets to show a profit. Money is made by totalling the assets gained from:

1. Land valued at the purchase price
2. Corn corps which are used to purchase land, seed, and water (Seeds sell for $5 a bushel, but corn sells for $25 a bushel. Corn crops make the most money, but are harder to obtain. The going rate of seed and corn is determined by the general store manager.)

3. Pelts and salt licks which have been discovered (These represent quick money, but are a non-renewable resource, so the supply diminishes as the year progresses.)
4. Materials purchased at the general store valued at their purchase price
5. Cash obtained from selling goods, pelts, salt, land, and water.

IV. Possible Problems

A. Some students may have a difficult time understanding the surveying and which areas are available. Adult leaders may need to offer assistance, perhaps for a price.
B. Some disagreements may occur over common passages through private land. The frontiersmen and settlers usually were generous with their neighbors and did not make them pay tolls to cross their land or use their water. If there is a draught, however, owners may refuse free privileges and place a toll on them.
C. If a family gets the idea to purchase land occupied by the general store, they may do so. Afterward, they may charge the general store rent.
D. Families may be in the process of surveying land that has already been purchased. The land officer has the right to refuse these claims or claims for overlapping land areas.
E. Any other problems which are taken to the land officer or can be discussed at the year-end town meeting.

V. First Town Meeting

A town meeting is called at the end of the first year. All families gather to review the previous year.

A. Discuss problems that have occurred in the settlement.

1. How do people feel about the land officer? The general store manager? The surveyor?
2. How were the relations with the highwaymen? Usually students are furious with the highwaymen who have cheated and lied.
3. What other problems have developed? What are the solutions? How did settlers solve their problems in 1820?

B. Families must prove they have obtained land, water, clothing, shelter, and food.
C. Families tally their assets on the tally sheet. (See Figure 5.32.) Who gained the most money? Who lost the most money?

1. Assets

a. Cash — Any actual cash money a family has in hand.
b. Equipment and supplies — The purchase value for any item bought at the general store or from another family, including surveying equipment.

c. Land value — The purchase price of any land bought.
d. Crop return — the value of corn harvested at the selling price per bushel

2. Tax and fines

a. Total taxes and fines according to the instructions.
b. Subtract that total from total assets to find the families' earnings for the first year.

D. Discuss what kinds of things each family might do differently.
E. The cards, representing consumed items, are collected. These include cards for clothing, seed, water, and food (at least one pelt or sheep card). Land deeds and equipment cards, such as animals, plows, wagons, and nails, are not collected.
F. The settlement representative then draws a chance card (or several at the discretion of the land office) which sets a new condition for next year. The second year then begins.

VI. The Second Year

The second year is also twenty (or thirty) minutes long. It is basically a continuation of the first with some possible differences.

B. Settlers may survey and purchase more land. If desired, they may purchase adjacent plots which are already claimed from one another.
C. Families may plant more crops, obtain more pelts and salt licks (if any are left), or sell goods to obtain more cash.
D. Highwaymen may become more helpful and honest, since they discovered settlers refuse to deal with them. After regaining various families' trust, however, they may start to cheat again.

VII. Second Town Meeting

A town meeting is called again at the year's end. This meeting serves as a wrap-up for the class.

A. Families fill out the tally sheets for the second year. (See Figure 5.33.)

1. All goods can be turned in at the general store for their purchase price. Then families simply count their money.
2. The second year's tally sheet differs slightly in the taxes and fines because the land and shelter requirement already has been fulfilled.

B. Discuss the problems encountered and the possible solutions.
C. What could have been done differently?
D. How did the second year differ from the first?
E. Was the class realistic? Did settlers run into some of the same problems?

Homesteading

Objectives:

1. To help students gain a view of pioneer lifestyle and some of the problems faced by homesteaders of the Northwest Territory in the early nineteenth century.
2. To acquaint students with some of the basic tools of the pioneer homestead.

Equipment:

1. Five to eight large sheets of paper and an equal number of felt markers
2. Possibly a few pioneer toys to demonstrate, such as the jumping jack, climbing bear, or the gee-haw-whimmy-diddle
3. Tools, as many as possible which are illustrated in the Pioneer Tools class (An effective class still could be taught with nothing but a felling axe.)
4. Some horse-drawn farm implements (These are not necessary, but add much to the class.) Implements might include a hayrake, hayloader, cultivator, and corn planter.

I. Introduction

A. Demonstrate the pioneer toys to catch students' interest and introduce the class topic.
B. Discuss the definition of pioneer. Define it as anyone who does something first. Are there pioneers today? The term *pioneer* usually refers to early Americans who settled the frontier areas. These frontier lines moved west as the years progressed. Originally, the frontier line was the Appalachian Mountains. Gradually, as that area settled, it extended to the Mississippi, then to the Rockies, and finally to the California coast. This class deals with pioneers of the Ohio and Michigan areas in the early 1800s.

II. On the Move

A. Students role-play settlers on the move from eastern Pennsylvania to southern Michigan, a move of about 600 miles.

1. Why would settlers want to move?

   a. Land was cheap. It often was $1 an acre or less, sometimes free. Land in the east might have been $25 to $50 an acre.
   b. People wanted freedom, excitement, and elbow room.
   c. Many early homesteaders were indentured servants and slaves who escaped into the wilderness.
   d. Fugitives from the law also bought land in the new territories because few questions were asked. (Simon Kenton, noted frontiersman, became a woodsman when he feared he had killed a man then fled to the west.)

2. What time of year would we want to leave for the west? Discuss advantages and disadvantages of each season.
3. What is the land like where we are going? Much of northern Ohio was swamp. Land that was not swamp was thick, unbroken forests full of wildlife (deer, squirrels, rabbits, bear) with many lakes and rivers throughout. In 1800 a squirrel probably could go from the Ohio River to Lake Erie or across the whole state of Michigan without touching the ground. Trees measured as much as forty feet in diameter.

B. Break students into groups of three to five and give each a large sheet of paper and a marker. Each group represents a wagon group. Give them ten minutes or so to write on their paper:

1. The time of year they wish to depart from Pennsylvania
2. Their estimated date of arrival
3. A list of ten items they are taking with them, in addition to their wagon and horses or oxen (Lists may include the list below.)

   a. Plow (Iron was not available upon arrival so metal parts of tools were taken. Wooden parts were made later.)
   b. Ax (This was an indispensable item used along the way and when pioneers arrived.)
   c. Guns and ammunition (These were used for hunting as well as protection.)
   d. Some staple food items such as sugar, flour, cornmeal, dried fruits, vegetables, and meat (Fresh meat was easily obtained all along the route.)
   e. Salt (This was for preserving meat more than flavoring.)
   f. Seeds for planting crops
   g. Farm animals (These may have included chickens, pigs, cattle.)
   h. Iron pots or skillet (Iron was not available away from settled areas and clay pots cracked and exploded in fire.)
   i. Medicines (Many were made as needed from herbs and plants.)
   j. Water (Perhaps just a bucket was kept full for stops between watering places which were plentiful.)
   k. Utensils, dishes, cups (These were usually made as needed from wood.)
   l. Spinning wheel, furniture, books, china (A few favorite family items may have had a place in the wagon, but were the first to be abandoned if the wagon had to be lightened or combined with another wagon.)

C. Discuss lists, possibly combining ideas to come up with one master list.

III. Arrival

A. Discuss the ideal place to homestead. What would one look for? A few important characteristics might be proximity to water, a river for transportation, rich soil, fairly flat ground for farming, not low or swampy ground, and not too many huge trees.

B. Each wagon now lists in order the jobs they would have to do on their new homestead site. Lists will vary, but a likely one might be:

   1. Clear the land. This involved chopping down trees, burning, pulling, and digging out stumps and removing rocks.

   2. Plant crops. If they arrived in the spring, this was more important than building a cabin since the family could continue to live in the wagon as they did on the journey. If they arrived too late to plant, they had to live on staples brought with them, hunted meat, and plants they could gather. Families often had a very lean first winter.

   3. Build a cabin. Often, the first cabin was little more than a lean-to shelter. Later, a single-room log cabin was built. It had a mud and rock chimney, a dirt floor, which was hard-packed and slept, and beds made of boughs or woven rope mattresses. Furniture might have consisted of a wooden table, wood stools, wood plates called trenches, candles, a cast-iron pot on the fire, and hooks on the wall for clothes. As years went by the log cabin was replaced or added onto with saw-mill cut planks; thus it became a frame house.

   4. Build a barn. Often the barn was built before the cabin while the family lived in the wagon or lean-to. The barn provided storage space for the crops and shelter for the animals.

   5. Work at ongoing tasks. When all this had been accomplished, there were more tasks: clearing land, moving stumps and rocks, hunting meat, planting, harvesting, and preserving food; making clothes, and many other constant chores.

IV. Tools

A. Students handle and examine tools, discussing what they are and how they are used. (For information, see Pioneer Tools class.) If only a felling ax is available, discuss all that it can do. Emphasize that a plot of land was cleared, a cabin built, fencing erected, and a barn built using only this tool.

B. If horse-drawn farming implements are available, have students examine them. Have them figure out their uses, where the horses were hitched, and where farmers walked or rode. Dates and names of manufacturing companies often can be found on the implements. Look for differences in the machines, early and later ones. Discuss when vehicle-drawn implements came into use: steam engines, tractors, combines, etc.

Challenge!

chapter 6

There are a number of activities in the following chapter that by their very nature involve risk. If an activity is approached without proper supervision or understanding on the part of the instructor, a student may be exposed to danger (the danger of bodily harm). It is always recommended that an instructor:

— further research the activity with an instructor or camp who is familiar with risk activities,
— plan the activity thoroughly,
— discuss thoroughly, the risk involved and the importance of group cooperation, with the students before each risk activity, allowing the students an option of participating or not,
— be aware and follow all regulations concerning safety that may apply,
— test the activity with instructors before involving students, and after having reviewed all information applying to the previously mentioned points.

Play it safe.

Orienteering

Objectives:

1. To help students understand the necessary concepts of distance and direction.
2. To help students understand the needs and possible uses of a compass.
3. To help students learn parts of a compass.
4. To help students understand that degrees are more accurate than the labels north, south, east, and west.
   To help students understand the concept of pace and how
5. to convert paces to meters.
6. To help students learn how to use a compass.

Equipment:

1. Compasses (This outline is based on the Silva compass, which is good for teaching.)
2. Model compass (See Figure 6.1.)
3. Ten-meter rope

During this hour class, students are taught how to use a compass and explains the purpose for using a compass. Pose various questions and have students come up with answers for the following information. Later they put this knowledge to practical use on an orienteering course.

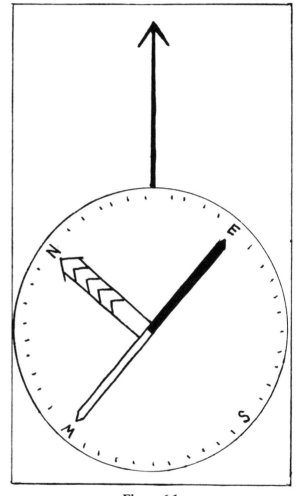

Figure 6.1

I. Introduction of Orienteering

A. Orienteering is considered a sport in Scandinavian countries.
B. It is increasingly recognized as a competitive sport, which is often combined with running or skiing.

II. Direction and Distance

A. These are two important concepts for anyone who wants to know which way and how far a certain place is.

1. The following may be used to find directions without the use of a compass.

   a. North star
   b. Moss growing on the north side of trees (Be careful because this is not always true.)
   c. Sun rising in the east and setting in the west
   d. Marking trails
   e. Marking landmarks
   f. Instructions, such as left or right

2. These may be used to measure distances.

   a. Yardstick, ruler, meter stick
   b. Use of body (hands, feet, arms, legs)
   c. Pacing (a normal step)

B. Necessary concepts for using the compass include:

   1. Direction — described in degrees
   2. Distance — measured by pacing and converting the individual's paces to meters

III. Compass Parts

A. *Compass Housing* — Within this house are the numbers (degrees) and the arrows.

   1. Each cardinal point is 90 degrees from the previous one.
   2. A complete circle is made up of 360 degrees.

B. *Red Needle* — The red needle ALWAYS points to magnetic north.

   1. Magnetic north is located in Canada.
   2. Magnetic north is different from geographic north.
   3. The difference between magnetic and geographic north (declination) must be considered.

It is important to tell students that they should never simply follow the red needle's direction because it always points north.

C. *Traveling Arrow* — This arrow gives the direction to travel.
D. *Red-Striped Arrow* — This arrow is located in the center of the compass housing. The traveling arrow points to the direction of travel after the red-striped arrow lines up with the magnetic needle.
E. *Degrees* — These are numbers on the compass which give accurate directions.

   1. Degrees are more accurate than the general directions of north, south, east, and west. Why?
   2. These are the same degrees used in longitude and latitude bearings.

IV. Holding a Compass

A. Hold a compass between two thumbs. Bring both arms into the body, positioning the elbows into your sides.

B. Hold the compass flat, parallel with the ground.
C. The traveling arrow should point straight ahead of you.

V. Working the Compass

A. Suggest a certain degree (direction) to go (30 degrees).
B. Turn the compass housing until 30 degrees lines up with the traveling arrow.
C. Turn your body, not the compass or dial, by taking small steps in a circle until the red-striped arrow and the magnetic needle line up. Remember to hold the compass steady. This method is important because it allows you to face the direction you want to travel. (See Figure 6.2.)
D. Let students use the compass. Make sure they understand the parts and the correct way to hold it.

VI. Activities and Exercises

A. *Stories* — Make up stories or adventures to get students excited about finding directions. An effective one is about a UFO landing in the area. After a rather lengthy and almost believable description of meeting Martians, students quickly try to find the direction of the landing.
B. *Picking an Object* — Have one student stand in a certain spot and find an object's direction. Then

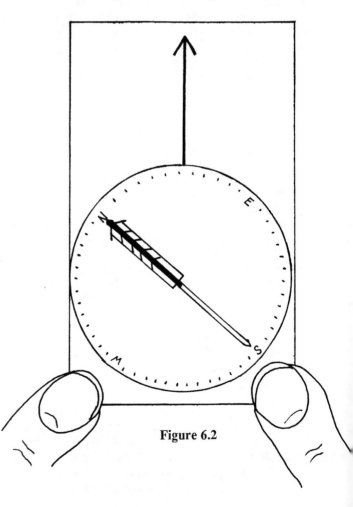

Figure 6.2

the student gives the compass to a partner and tells the direction and position to stand. The partner figures out the object.

C. *Making Squares and Triangles* —Have students mark and start at one point.

    1. To make a square, give the first bearing then add 90 degrees to each of the next three times, keeping the same pace all four times. Change pace with the second and fourth instructions to make a rectangle.

    2. To make a triangle, follow the same procedure but add 120 degrees each time, keeping the same pace all three times.

VII. Pacing Distances

A. Everyone's pace (normal step) is different. Why? Have students walk alongside a ten-meter rope several times to find out how many steps it takes them to go ten meters.

B. Involving students' math knowledge, figure from an average pace the amount of steps it would take to go 20 meters, 35 meters, 50 meters, etc., until all understand.

C. A rough estimate of one and a half paces per meter usually works well. To convert, students can multiply meters by 1 then by one-half and add the two resulting figures.

    Example:

$$1 \ 1/2 \text{ paces x } 16 \text{ meters} =$$
$$(1 \times 16) + (1/2 \times 16) =$$
$$16 + 8 = 24$$

VIII. Alternative Activities

A. Students can blindfold themselves and walk to a certain object. By placing a stake every ten steps, they can see their paths and realize how difficult it is to walk in a straight line without a compass.

B. Make a small course and let students practice finding directions and figuring paces.

C. Students can make their own orienteering course.

D. Students can spin around ten times with their eyes close and try to orient themselves through the use of another student's voice.

E. Students search for and rescue a victim by following an orienteering course and using first aid abilities. (See Search and Rescue class.)

F. Map and scale down to size a specified area and find its objects.

G. Students can learn triangulation, the use of a compass to find the distance of a far-away object from two focal points.

Incredible Journey

Objectives:

1. To improve students' abilities to constructively work with a small group toward a common goal.
2. To improve students' use of problem-solving techniques.

3. To give students an opportunity to practice using the compass.

I. The Class

The Incredible Journey is a course consisting of a card collection of obstacles and tasks which can be overcome (or accomplished) only by the combined efforts of a small group. It may incorporate compass use which students learned in the Orienteering or Survival classes. A course may consist of four to seven challenges which are scattered in a large area. Therefore, it is important that cards include straight bearings and paces to each challenge if compasses are used.

A group of twenty students is broken into three small groups, each led by a staff person, teacher, and/or other leaders. After an introduction, groups meet with their leaders and begin their courses, which can be started at different points. For example, one group begins with Challenge 1, the second with Challenge 3, and the third with Challenge 6. Groups then rotate through the course, taking challenges in order.

II. The Course

Following are sample challenges. Many others also can be invented with some creative thinking and a little research. Illustrations are not on the cards, but are included with this text to help explain the challenges.

CHALLENGE 1   Go 70 degrees for 80 meters.

> *This stone served as a makeshift dock,*
> *The pilgrims called it Plymouth Rock,*
> *Now all get on and that's no jive,*
> *You must stay on 'til the count of five*
> *without touching the ground, you clowns!*

(See Figure 6.3.)

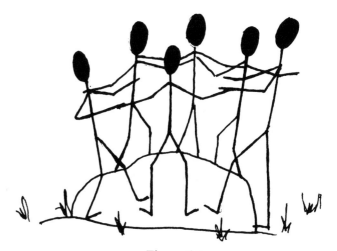

Figure 6.3

CHALLENGE 2 Go 108 degrees for 36 meters.

When you fall,
You must trust all.

Each member (paratrooper) in turn, must stand rigid on the airplane (picnic) table and fall backward from an upright position, trusting that the parachute (teammates) stops his freefall. Each paratrooper depends on his teammates to work together and form a soft landing spot. (See Figure 6.4.)

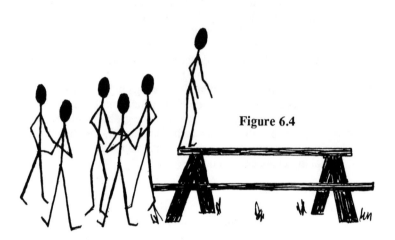

Figure 6.4

CHALLENGE 3 Go 310 degrees for 50 meters.

Cross this river using 2 x 4's,
If you drop the board,
You'll have it no more.

Using 2 x 4's, your team tries to cross the river. If either board even touches the ground, it is swept away by the current and can be used no more. A team member also is out if he falls and the rest of your team must go on. (See Figure 6.5.)

CHALLENGE 4 Go 40 degrees for 44 meters.

Get on this log and number off,
Then reverse your order without falling off.

Your whole team must line up on a log and number off beginning with 1. Then, without touching the ground, reverse the order of numbers with 1 at the other end. (See Figure 6.6.)

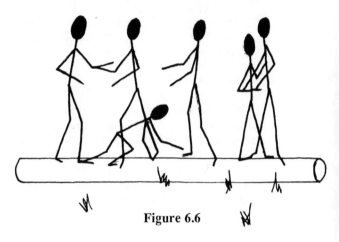

Figure 6.6

CHALLENGE 5 Go 326 degrees for 100 meters.

This rope is a poisonous vine,
Go over without touching,
And you'll be fine.

Use teamwork to get your whole team over the rope. You cannot climb or use the trees that hold the rope. You also cannot high jump the rope. (See Figure 6.7.)

Figure 6.5

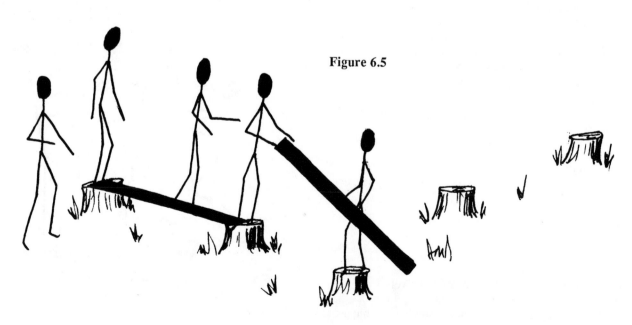

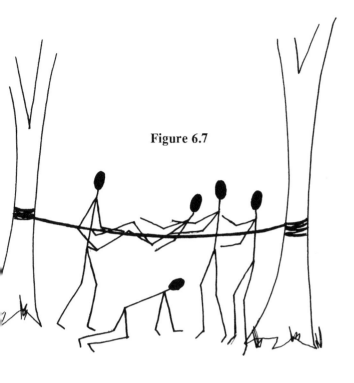

Figure 6.7

CHALLENGE 6   Go 210 degrees for 54 meters.

Disarm this atomic bomb,
And all will remain calm.

Your team of demolition experts must remove the explosive device (the tire) from the bomb (the post) to save the world. When finished, work together to replace the tire on the post for the next group. (See Figure 6.8.)

III. Other Suggested Challenges

A. *Skis* — Two long 4-inch by 4-inch boards each have eight short ropes coming through holes drilled through them. Team members must place the boards parallel to one another and stand with a foot on each. Holding a rope from each board, members must work together to walk from one spot to another. (See Figure 6.9.)

B. *Grand Canyon* — A log is suspended between two trees which have two long ropes hanging from them. Using only the ropes for support, team members must walk from one end of the swinging log to the other. If they fall off, they perish. (See Figure 6.10.)

C. *Teeter Totter* — A 2-inch by 8-inch board about ten feet long is placed on a stump or short log which is lying sideways on the ground. The team must divide so members are balanced on each end of the board in an attempt to balance it over the stump for at least five seconds. (See Figure 6.11.)

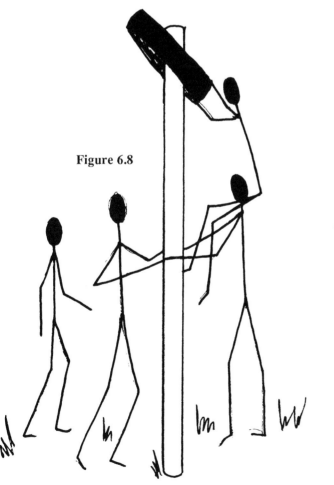

Figure 6.8

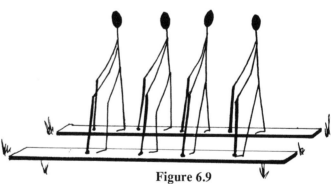

Figure 6.9

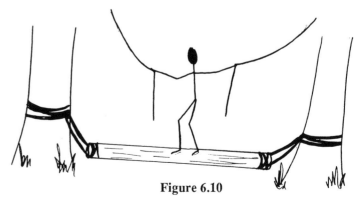

Figure 6.10

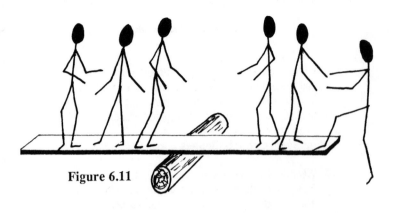

Figure 6.11

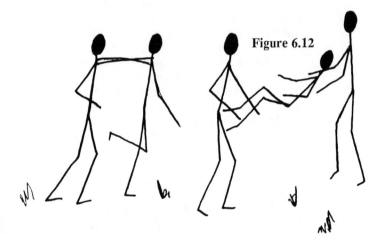

Figure 6.12

D. *Ambush* — All team members have been injured. Several are blind, some cannot use their legs, some lost the use of their arms, and several lost their voices. Working together, the team must travel a certain distance. (See Figure 6.12.)

E. *Medicine Bag* — Members, in turn, become blind and must be directed by voice only through an area of trees to a medicine bag that they can wipe across their eyes to restore sight. Team members who direct must stand only at the start of the path and cannot walk with the blind person. One by one, each is directed to the medicine bag until the entire team is reassembled and each person's sight is restored. (See Figure 6.13.)

F. *Stretcher* — Two logs are tied V-shaped between two trees. One pair of team members, starting at the closed part of the V, try to walk down the logs. Depending upon each other for support, they place their palms together and strive to reach the open end of the V. (See Figure 6.14.)

Mission Impossible

Objectives:

1. To encourage students to work together on various challenges which involve initiative and problem-solving techniques.
2. To make students aware of the group process: the intricacies of striving toward a common goal, and various roles that are unconsciously played by the members.

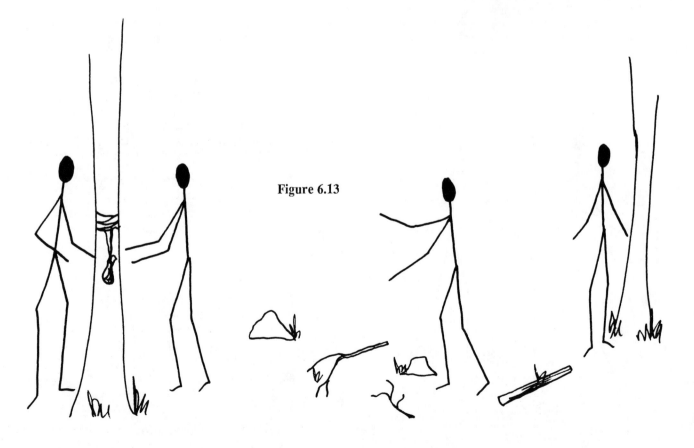

Figure 6.13

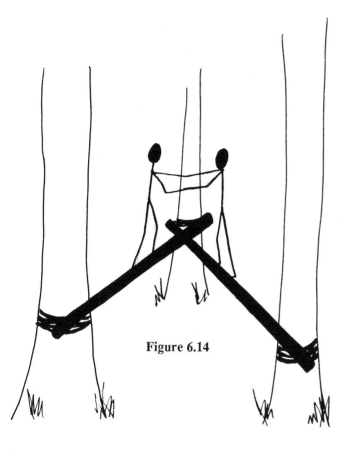

Figure 6.14

This two-hour class is similar to the Incredible Journey and Survival because it involves challenges which need to be accomplished by a group working together. It is different, however, because it is geared toward older students and forces them to examine the process used to solve a group's problems. The important goal of the class involves not the challenges, but the discoveries group members make concerning their interactions.

I. Introduction

   A. Discuss group cooperation to get students thinking about group work.

   B. Present a short group initiative challenge for a sample group to solve while other students watch. Thus, some observations are made concerning the group's problem-solving process.

   C. Play a tune, similar to the ''Mission: Impossible'' theme to explain the class and present the challenge. The mission is to survive the next two hours of challenges.

II. The Compass

   A. Review how to use the compass. Students should already know this skill; therefore, instruction should be fairly simple and brief. See the Orienteering class for details.

   B. Students should have a few minutes to practice the compass work, following and taking some bearings, until the instructor is confident that they understand how to use the compass.

III. Beginning the Course

   A. Instruct students to divide into three groups with approximately the same number of boys and girls in each. This dividing process is part of the class which should be discussed later.

   B. Each group gets a bag containing the necessary equipment for the course: the packet of instruction cards, one compass, blindfolds, and anything else needed for the various challenges.

   C. One leader, acting only as a silent supervisor, accompanies each group. Leaders take their groups to the starting points: from there students are on their own. They must read the directions, find the challenge, and work it out by themselves.

IV. The Course

   A. See Incredible Journey for sample challenges. Any type can be used; however, challenges which only a group can solve are ideal. Provide enough to keep an average group busy for about two hours.

   B. Each challenge should begin with a bearing and distance from the previous challenge. Bearings and distances can be increased in difficulty to challenge more competent groups.

V. Conclusion

This class discussion is very important since the purpose of the course involves more than simply completing challenges. It can follow a few simple questions.

   A. Did your group have a leader? Did you choose the leader or did he surface accidentally and unconsciously?

      1. Did anyone else strive to be the leader?
      2. How did the group react to the leader?

   B. Did the group work well together? What were some of the problems your group had? How could your group have organized itself better?

   C. Was the first challenge hard? Was communication difficult? Were things unorganized and confused? Did succeeding challenges become easier?

   D. What challenge was easiest? Which was the hardest? Which was the most fun?

Survival

Objectives:

1. To encourage students to use problem-solving techniques so they can constructively work together.
2. To make students more aware of basic survival needs.
3. To encourage students to consider the importance of understanding survival techniques and that survival situations are not uncommon, but usually unexpected.

I. Introduction

   A. Share some real-life survival situations which students have heard or read. Discuss how the situations came about and how the people involved survived or why they did not.

B. Discuss the basic needs for survival.

1. *Food* — Where is food obtained in a wilderness situation? Point out that many plants are poisonous and even many edible ones cannot give enough energy to replace that which is lost while looking for them. Also point out that it may sound easy to fashion a weapon or trap to kill or snare an animal, but it is not easy to do.

2. *Water* — Where is water obtained? Discuss water purification, melting snow or ice, and digging for water.

3. *Maintaining Body Heat* —Discuss the fact that humans are warm-blooded; therefore, their body heat must be kept within a few degrees of 98.6 in order to survive.

   a. *Fire* — Discuss how to build a fire, its purposes, and the importance of good tinder.

   b. *Shelter* — Why is shelter important? Discuss how different types of shelters are affected by wind chill and the weakening effect rain can have on the body, even in warm weather.

   c. *Clothing* — What types of clothing are best? Discuss wool, a windbreaker, and the importance of waterproof clothing.

Which of the basic needs is most important? We can live for weeks with no food, only a few days with no water, and possibly only hours with no shelter during the worst of weather and indefinitely during the best.

II. Group Activities

Break students into three groups of five to seven for several introductory activities designed to stress teamwork. Each group is to remain as a team for the survival course.

A. All group members empty their pockets into a middle pile. Next, the groups look at each object and discuss how it could be used in a survival situation.

B. To prepare for Common Squares an activity that stresses communication and teamwork, cut out six large paper squares, cutting each square into three odd-shaped pieces. Then place all eighteen pieces into an envelope. Thus, each team should get an envelope containing six squares cut into eighteen pieces. Working together, each group must put together the six squares without talking. Students cannot talk or signal for pieces, but must wait for someone to notice this need and offer a piece.

C. Other comparable games or activities emphasizing teamwork also would be appropriate.

D. Teach students how to use the compass and pace off meters. Next, have students pair off and practice using these skills outdoors while you direct them.

E. Read the first card for the course, then give each survival team a box containing twenty or so items which were "salvaged" from a plane crash. Tell them that they can carry only six items from the box and they must work together to choose the ones they think are most important. The boxes could contain the following items:

1. Compass	11. Jar of syrup
2. Matches	12. Bottle of "rum"
3. Large tin can	13. Piece of rope
4. Tarp or poncho	14. Candy bar
5. Fishline and hook	15. Knife
6. Mirror	16. Signal flare
7. Whistle	17. Bible
8. Newspaper	18. Aluminum foil
9. First aid kit	19. Clothes pins
10. Insect repellent	20. Toilet paper

Groups must choose the first four items. If not, they must select and return their chosen items and get those which were omitted. They also lose two minutes from their first fifteen-minute day. Groups can choose any other two items to complete their selection.

III. The Course

A. At this point explain timing. Loss of time represents the group's weakening chances for survival. There is no way they can build back time once it is lost. The first day is worth fifteen minutes. For example, they get lost (more than five paces off course) and lose two minutes. This means they then have thirteen minutes to build a fire. If they cannot, they lose five more minutes, which gives them eight minutes to build a shelter during the next day. If the shelter is built poorly, they may lose one, two, or more minutes which means the third day may consist of, perhaps, seven minutes. Within the third day's time limit, they must signal a rescue plane to survive. If they cannot do it within the time frame, they die.

B. Divide a class of twenty students into three groups of six or seven. Group leaders take the groups on three separate courses at the same time. It is important that leaders are shown the course prior to class. Their role is simply to read instructions on the Survival Cards, quietly supervise students, and act as timekeeper. They should not assist the group in any way or answer any questions.

C. Three stopping points which represent the end of a day are on each course. Students find each one by using a compass: the first where they must build a fire and boil water; the second where they must build a shelter; and the third an open area where they must signal a rescue plan.

IV. The Cards

Read the first card. After groups choose the items they wish to carry, leaders take their groups to the starting points. Each of the three groups has a separate course to follow, so there are three different starting points. At the starting points, group leaders read their cards for the first day's journey.

When arriving at each destination (day end), a group leader reads the card for that day. Only the first paragraph of the card is read, not the section labelled "Notes for Leader". The number of minutes which fill the blank on the card depends on the group's previous performance, beginning with the proper selection of the six items they are carrying.

Below are sample cards for the course:

Survival

Your plane has just crashed. Your leader, the only one injured, has a slight bump on the head. Because of this injury, your leader cannot make *any* decisions.

You know your exact position. You are 20 miles off courses. Your radio went out, so you cannot send a message for help; but you know the compass bearings to get back to the original course where a search plan may be looking for you.

Your group must select from these items six things to take along on your journey. You have _____ minutes.

First Day's Journey

From Red Triangle: Go 290 degrees for 38 meters.

You have been walking all day without a drink of water. Evening is approaching and the temperature is falling. Your group is cold and thirsty. Together you must build a fire to keep warm and boil half a can of water for safe drinking. You have _____ minutes.

Notes for Leader:

— Destination: fire circle.
— Minutes only decrease or stay the same; they never increase.
— If the group goes five steps to the right or left of the circle, they lose two minutes because of the large hills.
— If the group does not build a fire, they lose five minutes.
— If the group builds a fire, but does not boil water, they are penalized two minutes.

Second Day's Journey

From Red Triangle: Go 210 degrees for 48 meters.

Night is rapidly approaching. The temperature is falling by the minute. A storm is approaching from the *northeast.* Your group must build a shelter in _____ minutes to survive the cold and stormy night.

Notes for Leader:

— Destination: red triangle on post.
— Minutes only decrease or stay the same; they never increase.
— If the group goes five steps to the left or right of the post, they lose two minutes because of swamps on either side.
— If the shelter is considered inadequate, assess penalties to attain stress goal.
— If the group does not build a shelter in the allotted time, they are penalized five minutes.
— If the group builds a shelter, but does not consider the storm direction (NE), they lose two minutes.

Third Day's Journey

From Red Triangle: Go 150 degrees for 46 meters.

A plan is visible on the north horizon. It will be in view of your group in one minute and stay in view for _____ minutes. During that time your group must signal the plane. If your group successfully signals it, you are rescued. If not, you die here, in the wilderness. Waving your arms or screaming is not adequate. Your time starts now.

Notes for Leader:

— Destination: field
— After the plane is signaled, please return to the classroom with *all* class materials and survivors. Wait for the other groups to return.

Choice

Leader reads or explains:

"When I was young, my father took me hiking in this area. It looks very familiar. I know the compass bearings will take us to our destination without risk of losing minutes. However, I seem to remember a shortcut, but I can't remember if we go left down the valley or take the stream bed to the right. One is the shortcut; the other is the long way around. It's been so many years, I can't remember which way is right."

The Choice

A. Play it safe. Take the compass bearings and go.
B. Take a chance: 1. Shortcut: Gain two minutes
                   2. Long way: Lose two minutes.

Risk: 50-50 chance

Notes for Leader:

— Regardless of the students' choice, read the same compass bearings for the next destination.
— If students choose the shortcut, the two possible results can be symbolized by a small object hidden in one hand. Have students work together to decide which hand holds the object. If they pick the correct hand, they succeed on the shortcut and gain two minutes. If they do not, they miss the shortcut and lose two minutes.

V. Summary

When the groups have finished, reassemble and discuss how students felt throughout the course, their biggest problems, and how they would have done if the situation were real.

Wilderness Rescue

Objectives:

1. To give students a chance to use the compass in a practical situation.
2. To have students work as a team to accomplish various tasks.
3. To let students put their wilderness, first aid, and rescue knowledge to work.

Equipment:

1. One compass per group
2. Matches
3. Two ropes
4. Two blankets
5. A long sturdy stick
6. A mirror
7. Instant soup
8. Flashlight
9. One hat
10. One large can or cooking pot
11. First aid kit
12. Fishline
13. Tarp
14. Flat narrow board
15. Several granola bars
16. Any other desired survival equipment

This two-hour class, designed for junior high students, may be done with younger students. It is similar to the Survival class, but incorporates first aid and rescue techniques. The first hour is spent reviewing compass work and teaching first aid and rescue for the wilderness. The second is spent on a course similar to that in the Survival and Incredible Journey classes.

I. Introductory Discussion

A. Begin with a brief introduction about search and rescue teams to set the mood and develop enthusiasm.
B. Review the students' previous compass work. It may be necessary to teach the compass if they have not had a compass class.
C. Discuss basic survival needs. (See Survival class.)

   1. Food
   2. Water
   3. Shelter
   4. Warmth (clothing, shelter, fire)

II. First Aid

Teach or discuss various first aid concepts and treatments: first check vital signs, do not move the victim — unless necessary. Stay calm and always treat for shock.

A. *Victim Not Breathing* — Students only explain and demonstrate the procedure for mouth-to-mouth resuscitation. They do not actually breathe into the victim.
B. *Bad Cut* — By hand, apply direct pressure with a clean part of a shirt or cloth. The wound should then be wrapped with some sort of bandage.
C. *Shock* — Using whatever is available, keep the victim warm. Raise the head if the victim's face is red or raise the legs if the victim is pale.
D. *Broken Arm or Leg* — Splint the limb with sticks, tying with vines, belts, or anything possible. Then elevate it.

III. Discussion of Wilderness Situations and Rescue Approaches

A. Lost person search
B. Hypothermia, causes and cures, shelter and fire building
C. Ice rescue
D. Signaling for help
E. Frostbite
F. Transportation of victims

IV. Organizing the Class

A. A base station is controlled by the instructor.

B. The instructor has six cards, each with a different compass bearing and distance, and a box containing all the survival equipment.

C. Students are divided into groups of five or six.

D. Each group is given a card containing the bearings and distance readings.

E. Groups follow the bearings. At their destination they can find cards with further instructions on a post or tree. These instructions call for rescue operations. Victims are role-played by other adult leaders.

F. Rescue teams may return at any point to the base station for supplies. A team may take only two items from the survival equipment for each rescue operation. These two can be exchanged for the next rescue attempt.

G. When groups finish their rescue operations (or at the end of twenty minutes), they must return to the base station for a new set of compass bearings, which leads them to another card and their next victim.

H. Each group eventually follows all six bearings to each of the stations.

I. At the discretion of the victim and instructor, each group receives points for their rescue techniques. Each rescue operation is worth a maximum number of points. The team receiving the most points is the best in the area. (For suggestions concerning scoring, see Leaders Notes and Point Sheet, Figure 6.15.)

V. The Cards

A. *Instruction Card* — You are members of a top-notch wilderness rescue team. Your task is to respond with both speed and skill to various emergency situations. Each mission starts at central headquarters and leads into the wilderness at a designated compass bearing. On this particular day, you face six different emergencies. Choose one of the bearings which leads to the first challenge. Your goal is to complete as many tasks as possible. If you cannot complete a task within twenty minutes, move on to the next.

B. *Card 1* — Night is rapidly approaching. Six people have run into a snowstorm. Unless you quickly warm them, they may not live through the night. You must warm them (a fire and hot drink are best) and bring them back to the rescue station. Points: 8 maximum.

C. *Card 2* — A group has radioed for help. They are lost and need shelter to protect them from the cold and strong northwestern winds. One person is suffering from frostbite on his left hand and cannot assist in building the shelter. Points: 6 maximum.

D. *Card 3* — Your helicopter has located a person who has fallen through the ice. While it lands at the rescue station, save and treat the person. The ice is not strong enough to hold anyone. Time is essential. Move quickly. Points: 6 maximum.

E. *Card 4* — You have found a person lying on the ground. His right leg is in a distorted position. It is also discolored and swollen. Treat and transport the victim back to the rescue station. Points: 6 maximum.

F. *Card 5* — You receive a report that a hiker in the area has the following conditions: slurred speech, clammy skin, and stumbling walk. Find, treat, and return the person to the rescue station. Points: 4 maximum.

G. *Card 6* — A person is reported missing. His last known location is very near this area. Find the person. (A small outline of a person drawn on white paper is hidden here.) Signal the helicopter to rescue you. One point is awarded for each different signal made up to three signals possible. Points: 6 maximum.

The Beast

Objectives:

1. To challenge students to work well as members of a team.
2. To improve students' skills of observation and communication.

Equipment:

1. Tinker toys (enough to allow each group to build a beast, which identically matches the master beast, plus numerous assorted pieces)
2. Several hundred kernels of corn, or other item representing gold nuggets or some type of money
3. Two rooms, curtains, or some method of shielding the master beast from students' view
4. A master beast built by the instructor prior to class

I. Goal

Each team, using teamwork and communication skills, builds a beast exactly like the master beast.

II. Introductory Discussion

A. What is teamwork?

B. Communication and cooperation are important in teamwork.

C. Some roles may not appear important in a team, but the final outcome depends upon each person's ability to fulfill a particular role.

III. Procedure

A. The instructor builds a master beast with tinker toys and places it out of students' sight before class. The master beast can be simple or complicated, depending upon the competency level of students.

B. Students are broken into teams of four.

C. An adult leader operates a store which handles building supplies that are priced ahead of time by the instructor.

D. Groups must purchase the correct items to build a beast matching the master beast. No items can be

Leaders Notes and Point Sheet

1. Fire and Hot Fluids

— Did they build the fire? (4 points)   Yes _____ No _____
— Did they boil water? (4 points)   Yes _____ No _____
— Did they warm you with anything other than the fire and warm fluids?   Yes _____ No _____
— Did anyone comfort, etc., you while the fire was being built?   Yes _____ No _____

Make certain that the fire is completely out and all litter is picked up.

2. Shelter Building

— Did they consider wind direction (1 point)   Yes _____ No _____
— Is the shelter secure against strong winds? (1 point)   Yes _____ No _____
— Did the shelter cover everyone's head? (1 point)   Yes _____ No _____
— Did they treat you for frostbite? (3 points)   Yes _____ No _____
  a) Immediate warming?
  b) No rubbing?
  c) Continuous warming?

3. Ice Rescue

— Did they rescue you? (3 points)   Yes _____ No _____
— Did anyone go down the hill into the ice/water? (1 point)   Yes _____ No _____
— Did they treat you for shock and warm you? (3 points)   Yes _____ No _____

Don't walk up the hill; ''be heavy'' for them.
Play the role (very cold, etc.) until you return to rescue station.

4. Fractured Leg

— Did they treat you for shock?   Yes _____ No _____
— Were you continually calmed and reassured?   Yes _____ No _____
— Was your leg immobilized before a move was attempted? (3 points)   Yes _____ No _____
— Were you transported safely and comfortably? (3 points)   Yes _____ No _____
— Did they *gently* determine where the break was?   Yes _____ No _____

Your leg is broken below the knee.
Allow yourself to be transported to the rescue station.

5. Hypothermia

— Did they warm you with a blanket?   Yes _____ No _____
— Did they warm you by wrapping you with another person?   Yes _____ No _____
— Did they offer you food and drink?   Yes _____ No _____

A total 4 points are possible.
Please do your best to simulate this situation.

6. Search and Rescue

— Did they find the missing person? (3 points)   Yes _____ No _____
— Did they signal in an open area?   Yes _____ No _____
— Were the signals large and well spaced?   Yes _____ No _____
— How many different signals did they use?   Yes _____ No _____

Waving arms and screaming will not work!
For each different signal made 1 point is given. (maximum 3 points)

Figure 6.15

returned for a refund or exchanged. If a group runs out of money, they can attempt to purchase necessary materials with other items, being as creative as possible. The store owner decides the purchase values.

E. The master beast cannot be seen by the groups striving to duplicate it.

F. Groups are given a time limit, usually thirty to forty minutes, to complete their beasts.

IV. Roles of Team Members

A. The Observer

1. The observer looks at the master beast and reports to the relayer all the information needed to build a duplicate beast.
2. Being the only person who looks at the master beast, the observer checks sizes, colors, shapes, and positions.
3. The observer cannot touch the beast or write any information.
4. This person cannot see the duplicate beast which is being built.
5. The observer speaks only with the relayer, not to any other team members.
6. The observer views the master beast as often as necessary.

B. The Relayer

1. The relayer tells the buyer what items to purchase. He also tells the builder how to construct the duplicate beast to match the master beast.
2. The relayer cannot look at the master beast, but gets the needed information from the observer.
3. The relayer may go to the observer for information as often as necessary.
4. The relayer cannot write any information.

5. The relayer may talk with other team members and is the only person who can speak with the observer.
6. The relayer can see the beast being built, but cannot touch any piece of material.

C. The Buyer

1. The buyer purchases necessary items from the store owner.
2. The buyer is the only one who speaks with the store owner or purchases building materials.
3. The buyer obtains needed information from the relayer concerning which items to buy.
4. The buyer cannot look at the master beast.
5. The buyer is the only one who handles money or negotiates with the store owner if the money runs out.

D. The Builder

1. The builder constructs the team's beast.
2. The builder gets instructions from the relayer and materials from the buyer.
3. The builder cannot look at the master beast or speak with the observer.

V. Concluding Discussion

A. Regroup with both beasts visible. Bring out the master beast and compare it with the groups' beast.
B. Share problems and frustrations.

1. What were the difficulties and problems in communication?
2. Which of the four team members communicated the most?
3. Which member had to be the most precise?
4. Which member had the most boring job?
5. What would have happened if any team member had quit?

The Good Earth

Future Score

Objectives:

1. To introduce students to various energy sources, their limitations and the need to supplement primary energy sources with secondary units.
2. To give students a chance to use creativity and teamwork skills in designing their own dwelling, utilizing their resources, and financially supporting themselves in a model situation.

Equipment:

1. Markers or crayons
2. One-inch graph paper, one piece for each group
3. Corn kernels, acorns, or some item to represent money
4. Situation cards, about twenty
5. Energy Source Chart (See Figure 7.1.)
6. Maintenance and Operating Rate Schedule (See Figure 7.2.)

For class effectiveness, the instructor must be aware of energy sources, current research in this field, energy use problems, politics involved, and environmental and economic issues involving energy. It is important that the instructor do adequate research regarding the material. This information is simply a method of organizing material for a class and challenging students with it.

I. Introductory Discussion

    A. List and discuss as many forms and sources of energy as possible.

        1. Nuclear fission
        2. Natural gas
        3. Hydroelectric
        4. Solar
        5. Wood
        6. Coal
        7. Geothermal
        8. Secondary recovery petroleum
        9. Petroleum
        10. Tidal
        11. Oil shale
        12. Nuclear fusion

    B. Discuss various problems with these types of energy.

        1. Renewability
        2. Cost of installation
        3. Cost of maintenance/operation
        4. Hazards
        5. Environmental impact

II. The Class

    A. Students are divided into families of four.
    B. Each family designs their estate and dwelling and charts or maps it onto graph paper. Give them about five minutes.
    C. Each family is assigned a type of energy from the Energy Source Chart to operate their estate. (See Figure 7.1.)
    D. At the end of each five-year period each family is billed for energy use.
    E. Each family's goal is to survive and stay financially solvent for twenty years without going bankrupt or losing their estate.

III. Roles of Leaders

    A. One adult leader role-plays the bill collector who charges families for energy use according to the Maintenance and Operating Rate Schedule. (See Figure 7.2.)
    B. The banker, an adult leader, passes out money to families at the beginning and at the end of every five-year period (their salaries of two money units). The banker also offers credit to families who run out of money. To extend credit, the banker repossesses a portion of a family's land by marking a large black X through any one-inch square on the family's graph-paper estate. It can be a square the family or banker chooses.

IV. The Procedure

    A. Divide the class into families of four.
    B. Assign each family an energy source from the Energy Source Chart. Be sure every type is represented by at least one family.
    C. Discuss the chart by columns.

ENERGY SOURCE CHART

Energy Source	Availability	Environmental Impact	Installation Cost (in money units)	Insulation Cost (in money units)
Nuclear	Diminishing, non-renewable	Possibly critical	10	2
Natural gas	Diminishing, non-renewable	Medium	4	2
Hydroelectric	Plentiful, but depends on location of rivers/dams	Medium	6	2
Solar	Plentiful, limited research and technology	Slight	16	2
Wood	Diminishing, renewable	Medium	2	2
Coal	Diminishing, non-renewable	Medium	4	2

Figure 7.1

MAINTENANCE AND OPERATION RATES
(in money units)

Energy Type	Now	+5	+10	+15	+20
Nuclear	8	7	6	5	5
Natural gas	4	6	9	12	15
Hydroelectric	6	6	8	9	10
Solar	3	4	2	2	2
Wood	4	6	9	11	13
Coal	4	6	10	14	18

Figure 7.2

1. Energy sources have different levels of availability.
2. Each has a different level of environmental impact, short-term and long-term.
3. Various energy forms have higher costs of installation than others.
4. Insulation costs are generally the same regardless of energy use.

D. The banker passes out twenty money units to each family at the beginning of their twenty-year period.
E. Two and one-half years end. Two situation cards are drawn for the entire group. These represent happenings in economics, resources, politics, and science and affect energy sources and costs.
F. The bill collector then collects any money units which situation cards require.
G. The first five-year period ends. The bill collector collects money units due from each family for their energy as shown on the Rate Schedule.

H. Another two and one-half years end, and two more situation cards are drawn. (Cards continue to be drawn every two and one-half years.)
I. At the end of each five-year period, the banker gives each family two money units, which represent an income for those five years.
J. If at any time a family runs out of money units, the banker can repossess a one-inch portion of the family's estate by marking it out with a black X.

V. Situation Cards

These cards have possible changes or problems written on them. They randomly are picked at the end of every two and one-half years. Information for the cards may include the following.

A. James Watt opens federal lands to lumbering. Subtract two money units from wood costs. Landowners also lose some wilderness and must subtract one square from their ideal estate.
B. New gas reserves are discovered in the Midwest. Those using natural gas subtract three money units from their next bill.
C. This was a cloudy year. Those with solar power add three money units to their next energy bill.
D. New technology in solar power cuts maintenance cost in half. From this point on all solar costs are reduced by one money unit. Those who wish to switch to solar power may do so at this time at half the installation cost.
E. There is a wood shortage. Wood must be used from local lands. Those with wood lots subtract four squares from their estate.
F. Acid rain causes Congress to pass tougher clean-air laws. Power companies now must use low sulphur coal. Add two money units to coal costs from now on.
G. Environmental Protection Agency blocks dam completions at selected sites across the United States. Those using hydroelectric power add two money units to their next bill.

H. Scientists discover an easy, non-poisonous way of disposing of nuclear waste. Those using nuclear power subtract three money units from their next bills.

I. Construction of new oil pipelines are delayed. Add two money units to cost of natural gas next year.

J. Miners strike for better working conditions. Coal users add two money units to their next bill.

K. Inflation hits hard. Subtract two money units from everyone's total.

L. Congress lifts pollution controls on coal power stations. This cuts coal costs in half. One square, however, is subtracted from all estates due to acid rain damage on the environment.

VI. Discussion

A. The instructor needs to discuss national problems in politics, science, environmental developments, and economics which affect everyone every two and one-half years throughout the twenty year period. For example:

1. The Great Depression
2. Stock Market Crash of 1929
3. Two world wars
4. Alaska Pipeline
5. Types of energy used throughout the twentieth century, how they changed from one decade to another.
6. Energy laws
7. The Teapot Dome Scandal
8. Gas wars and OPEC

B. The final discussion at the end of twenty years, should deal with why everyone lost land and money.

1. What type of energy was best?
2. What would you change?
3. Where can we develop or invent other types of energy?
4. What form disappeared first within the twenty years?
5. Which types of energy have been improved or developed in the twenty years?
6. We, as consumers, need to consider all these issues.

Weather Mapping

Objectives:

1. To help students use observation skills in reading simple weather instruments.
2. To help students draw conclusions by using experimental data and a physical map.
3. To help students form and test hypotheses concerning physical influences on weather in a small area.

Equipment:

1. Five anemometers (See Figure 7.3.)
2. Five feather vanes or yarn tell-tales (See Figure 7.4.)
3. Five standard thermometers
4. Data sheets
5. Grid sheets
6. Map of the area on clear plastic
7. Five sets of crayons
8. Twenty-five stakes about twenty-four inches in length.

I. Preparation for Class

An experimental area must be staked off and mapped before class. It should be twenty-five meters square,

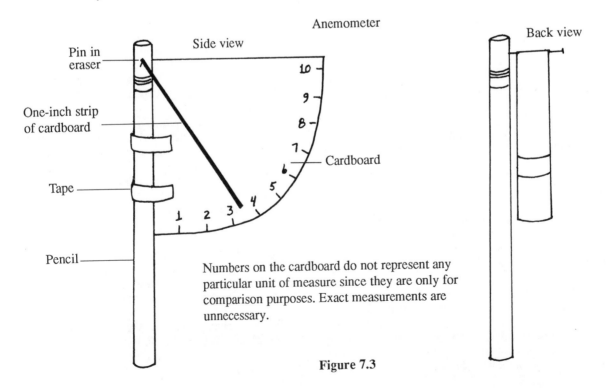

Anemometer

Side view

Back view

Pin in eraser

One-inch strip of cardboard

Tape

Pencil

Cardboard

10
9
8
7
6
5
4
3
2
1

Numbers on the cardboard do not represent any particular unit of measure since they are only for comparison purposes. Exact measurements are unnecessary.

Figure 7.3

Wind vanes

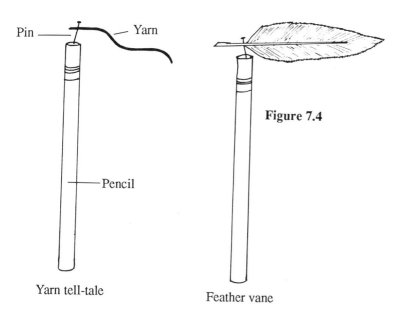

Figure 7.4

Yarn tell-tale

Feather vane

marked off with twenty-five stakes, each five meters apart. (An area with varying physical features works best.) A map indicating the major features of the area (trees, bushes, ditches, rocks, etc.) then should be drawn on clear plastic. This works as an overlay on recorded data.

II. Student Measurements

A. Wind Direction

1. Stakes should be lined up in N-S-E-W lines. These directions also should be pointed out to students.
2. Use the feathervane or tell-tale to find wind directions.

B. Wind Speed

1. Make certain the anemometer is facing the correct wind direction.
2. Measure wind speed at ground level and at a height of one and a half meters.

C. Temperature

1. First check thermometers in a controlled environment. They can vary as much as two degrees, so note differences and compensate when recording data.
2. Keep thermometers outside so they do not take long to readjust in the new area.
3. Hold thermometers so body heat does not affect them.
4. Allow one minute for thermometers to adjust to each new measurement.
5. Measure the temperature at ground level and at one and a half meters high.

III. Typical Class

A. Introduction

1. What is the difference between weather and climate?
2. Do you think the weather is any different between here and San Francisco?
3. Do you think there is any difference between here and Chicago?
4. Is there any difference between here and your home town?
5. Is there a difference between here and a mile away?
6. Is there a difference between here and a hundred yards away?

B. The instructor then explains how to read various weather instruments and the organization of the experiments. The class is divided into five groups, each getting a row of five stations in the experimental area to measure. Students measure wind direction, wind speed, and temperature.

C. Students remain in groups of five after data is gathered. Each group takes charge of a specific measurement, such as wind speed at one and a half meters, and transfers the measurement from a data sheet to the Data Grid Sheet. (See Figure 7.5.) When a data grid is complete, figures are coded with a colored X in the square. The progression might be:

	Temperature	Speed	Directions
Purple	Lowest	Slowest	NW or North
Blue			NE
Green			SE & East
Yellow			South
Orange			SW
Red	Highest	Fastest	West

In the case of wind direction, an arrow is drawn in the square going the same way as the wind, pointing toward the direction the wind is blowing.

D. When completed, the map overlay is placed over the data grid, and patterns and discrepancies are picked out. (See Figure 7.6.) Questions should be asked concerning how weather was specifically affected (for instance, the wind pattern around a tree or a temperature drop, or rise, in the path). Students often ask and answer their own questions.

IV. Alternative Activities

A. Instead of measuring in a specific pattern, measurements are taken in widely different environments and conclusions drawn as to the differences.
B. Map a wind pattern around an object (hill, building, tree) or negative object (ditch or hole).
C. Try a challenge. Who can find the most varied temperatures in your classroom?

WEATHER MAPPING DATA GRID

NORTH

A	B	C	D	E
F	G	H	I	J

WEST

| | | | | | EAST |

K	L	M	N	O
P	Q	R	S	T
U	V	W	X	Y

SOUTH

Type of Reading _____

(Include temperature at ground level, temperature at 1.5 meters, wind speed at ground level, wind speed at 1.5 meters, and wind direction.)

Figure 7.5

A SAMPLE OVERLAY WITH STATIONS

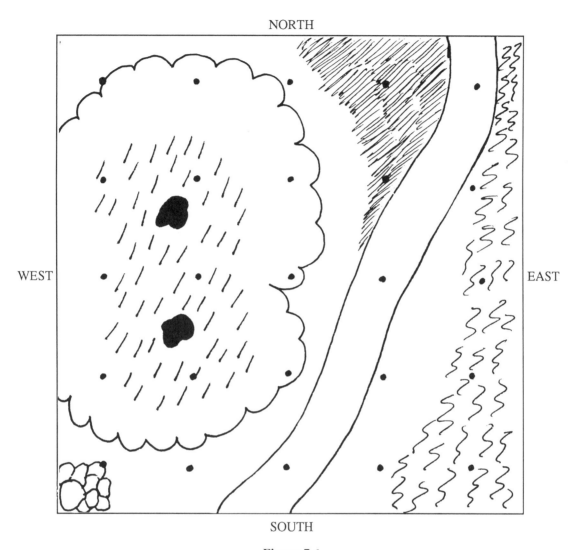

Figure 7.6

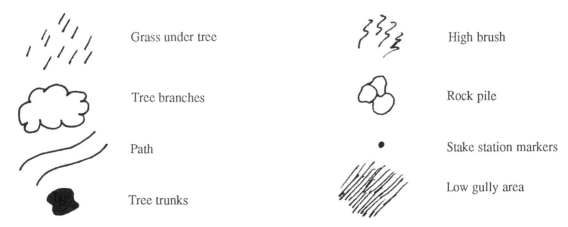

D. Send out helium-filled weather balloons with postcards attached. Use a clinometer to measure how fast the balloons rise and move from the launching site. Release balloons on days having different wind directions. Make a map showing where the postcards end up.

Front:

```
┌─────────────────────────────────────┐
│                            ┌──────┐  │
│                            │      │  │
│                            └──────┘  │
│                                      │
│                                      │
│             School Address           │
│                                      │
│                                      │
│                                      │
└─────────────────────────────────────┘
```

Back:

```
┌─────────────────────────────────────┐
│                                      │
│  We are a _____ grade class at ___  │
│  School. Please fill out and mail.   │
│                                      │
│  Thank you.                          │
│                                      │
│  Name _____ Age _____  │
│                                      │
│  Address _____ │
│                                      │
│  Place found _____ │
│                                      │
└─────────────────────────────────────┘
```

Soil Survey

Objectives:

1. To make students more aware of soil and its importance in sustaining life.
2. To teach students what soil factors must be considered before growing plants.
3. To help students use the scientific method in testing soil conditions and analyzing results.
4. To help students understand how a person uses soil testing equipment.

Equipment:

1. Commercial soil testing kits
2. Aluminum pie pans
3. No. 10 cans with both ends cut out
4. Two-pint measuring container
5. Thermometers

I. Introduction

A. What is soil? Would there be plants without soil? Animals? How is the soil formed? Do rocks become soil? What happens to plants when they die? Is all soil alike?
B. How does a farmer know if his land is good? How does he know where to plant which kinds of crops?

II. Horizons (Layers) of Soil

Take students to an area to be sampled.

A. Topsoil contains several types of materials.

1. *Litter* — What is it called when people throw trash on the ground? Dead plants and sticks also can be called litter. Pick up a handful and describe how it feels. List objects found in it. Using a ruler or a stick, try measuring its depth.
2. *Duff* — This is litter that has begun to break down. Is there any? If not, why? Is the rate of decay related to the amount of duff?
3. *Humus* — This is the rich, black material found under the duff. How is humus formed? How does litter decay? Describe and measure this layer. Where does the topsoil end?

B. Subsoil is the next layer. How can you tell where it is? Is subsoil lighter or darker than topsoil? Why? Is there less organic material than topsoil?
C. Dig down to the parent material if you can. Would it be harder or softer than subsoil? Why is it called parent material?

III. Tests

A. Temperature

1. Gently push the thermometer into the soil. Wait until the indicator stops before reading it. Next take the air temperature.
2. Is the soil warmer or colder than the air? Why? How will this affect animals living in the soil? How will it affect plants? Can the soil be too cold for things to grow? What is growing season? Discuss the *frostline*, the point below which the soil does not freeze. How does frostline affect where water pipes are placed? How does it affect plant roots and animals?

B. pH Test

1. A commercial soil testing kit is needed. Several types are available. Discuss sweet and sour tastes. Use this concept to explain the pH of soil, acid and alkaline.
2. Explain that different plants grow in soils with differing pH. Place some soil in a pie pan and add several drops of indicator liquid. Mix well and match the color to the test chart.

3. What plants grow best at that pH? What grows in the soil where the sample was taken? What would happen if you took a plant growing in soil with pH 5.0 and moved it to soil with pH 8.0?

C. Color

1. What color indicates good soil? Why?
2. What does fertilizer add? How did early Americans fertilize the soil? Manure and fish are two types of fertilizer which become a natural humus.
3. Which is more fertile, topsoil or subsoil? What happens to eroded soil? What color is it?

D. Texture

1. If possible, have samples of clay, sand and silt. Have students feel each and discuss textures.
2. Why is texture important? Does water drain through some textures faster than others? Is air space important? Can roots push more easily through one texture than another?

E. Permeability

1. Push a No. 10 can with both ends cut out into the ground. Quickly pour about two pints of water into it. Measure the length of time it takes the soil to absorb the water.
2. Define permeability. Does permeability relate to texture? What other things affect permeability? (Other factors are looseness of the soil, moisture, ground cover, and litter.)

F. Results

What type of soil is best for farming? Do farmers want all water to drain out? Do they want soil that holds so much water that puddles stand? Is the soil in this sample area good for farming, a road, a septic tank?

IV. Activity

A. Divide into three groups. Each group tests a different area such as a pine forest, a deciduous forest, and a field.
B. Each group records its data on the Soil Survey Worksheet. (See Figure 7.7.)
C. When finished, regroup and compare the three areas. Which would be suitable for farming? Which land would each group like to buy?

SOIL SURVEY WORKSHEET
Figure 7.7

I. Soil Horizons or Layers

A. Topsoil

1. Litter (dead material found on the surface which is clearly identifiable). List objects found:

Description (feel, appearance):

Depth _____

2. Duff (litter that has begun to break down) Where is this layer found? Why?

Description:

Depth _____

3. Humus (completely decomposed litter, rich black material)
Description:

Depth _____

B. Subsoil (contains little organic matter)
Where does it begin?
How far did you dig?
Did you dig past the subsoil?
Describe:

C. Parent material
Can you find it?
Description (guess if you do not find it):

II. Testing Conditions

A. Temperature: Air _____
Soil _____
Which is warmer?   Why?

B. pH — Circle the pH on your plot.

Acid   1-3   too acid for most plants
                              4
                                           blueberries, ferns, camellias
                              5
                                           pine, fir spruce, oak, birch, willow
                              6
                                           maple, peach, carrots, lettuce, pine
Neutral   7   most plants do best here
                                           beech, asparagus, cinquefoil
                              8
                              9
Alkaline   10-14   too alkaline for most plants

Which plants grow best here?

C. Color of your soil: _____

Using the chart, determine fertility and erosion:

Figure 7.7 (continued)

Color:	dark brown or gray to black	brown to yellow black	pale brown to yellow
Fertility	Low	Medium	High
Erosion Level	Excellent	Good	Low

D. Texture (rub some between your fingers):

If it feels gritty . . . . Sand
If it feels smooth and slick . . . . Silt
If it feels very smooth and sticky . . . . Clay

E. Permeability:

Time until water is gone _____
Is this affected by texture?

F. Texture chart: Where does your soil fit?

Texture	Ability to Hold Water	Looseness
Sand	Poor	Good
Silt	Good to Excellent	Good
Clay	High	Poor

III. Evaluate Your Plot

Judging from thickness of topsoil and color, how fertile is this area?

What plants can grow in this pH?

Are the texture and moisture suited for planting crops?

Any other considerations?

Will you decide to farm here?

Figure 7.7 (continued)

V. Other Activities

A. Compare soil conditions in various areas. For example:

1. A trampled path and an unused area nearby
2. A horse pasture and an open field
3. A sloping area and a low, flat area

B. Collect soil samples from various areas and attempt to construct a key comparing texture, color, and pH.
C. Have a high-low hunt. Small teams go out and try to find soil with:

1. The highest and lowest pH
2. The highest and lowest slope
3. The fastest and slowest permeability

D. Do an erosion study. Map an eroded area to determine where erosion occurs. Splash sticks to see where rain hits soil the most, and determine the slope.

VI. Post-experience Activities

A. Make a compost pile and observe changes. Make two, perhaps, adding lime to one and not the other. Then compare the pH.
B. Mulch around tree roots. Collect and use old Christmas trees to mulch a nature trail.
C. Compare the sprouting and growth of seeds in a variety of soils.
D. Make a worm farm sandwich in a glass. Use various layers such as light sand, cornmeal, coffee grounds and rich humus to see how worms move soil and create air space.
E. Collect different types of dead leaves and put them in a damp, dark box with several sow bugs. Record which leaves disappear at what time.
F. Make settling jars. Mix soil samples with water and shake them up in a glass jar. Watch the soils sort out by particle size as larger pieces settle first.
G. Find an eroded area and build check dams or plant grass to stop soil from washing away. Compare effectiveness of the two methods.
H. Try sand painting. American Natives traveled far and wide to collect different colors of sand which they used to create temporary pictures in special ceremonies. After different colors have been collected, glue can be used with construction paper to make permanent pictures.
I. Use a light bulb and a bleach bottle with the bottom out to make a Berlese funnel. Use it to find out what animals are in the soil. Place soil on a screen over the small opening. As the soil heats, insects crawl toward the opening, away from the light, and into a jar with water.

Mapping

Objectives:

1. To teach students to identify the important aspects of a map.
2. To encourage students to utilize their knowledge of the compass in mapmaking.
3. To help students understand the idea of a grid in mapmaking.

Equipment:

1. Four compasses
2. Ten identical state road maps
3. Graph paper
4. Eight ropes, each ten meters long, marked with yarn or tape at one-meter lengths
5. Pencils
6. Clipboards or other hard writing surface
7. A five-meter square grid marked on a classroom floor using chalk or tape with meters and half-meters marked off along the edges and sides oriented along the lines of the cardinal directions.

I. Introduction

A. Hand out state maps. Discuss the fact that maps are scaled-down pictures of a real area.
B. Discuss the legend on maps: directional symbols, scale of miles, and other symbols.
C. Discuss the grid, noting the cross lines indicated on the margins by letters and numbers. Have students locate several towns by using keys and grids.

II. Explanation of Mapping

A. Call students' attention to the five-meter-square area marked on the floor, representing an area of land.
B. Drop an object into the area. Ask students to figure out how they could draw the exact location of the object on a piece of paper. Discuss the problem of accuracy.
C. Pass out the graph paper. Have students figure out how the graph paper should represent the land area marked on the floor. Figure a scale of meters. (Example, one large square on the graph paper represents one meter.) Use whatever scale is practical for the type of graph paper and the students' comprehension level.
D. Have students label the marks on the borders of the floor area and on the graph paper with letters and numbers so they have a grid similar to that found on the state road map.
E. Call attention to the fact that land area borders follow a north-south and east-west orientation. The task is to find out exactly where the object lies on the east-west line and along the north-south line.

1. To find the object's placement along the north-south line, have a student (surveyor) stand on the east side of the area with a compass set for due west. The student lines up the traveling arrow with west and, sighting along the arrow, moves along the east boundary line until the arrow is pointing directly at the object. The student's position then is charted along

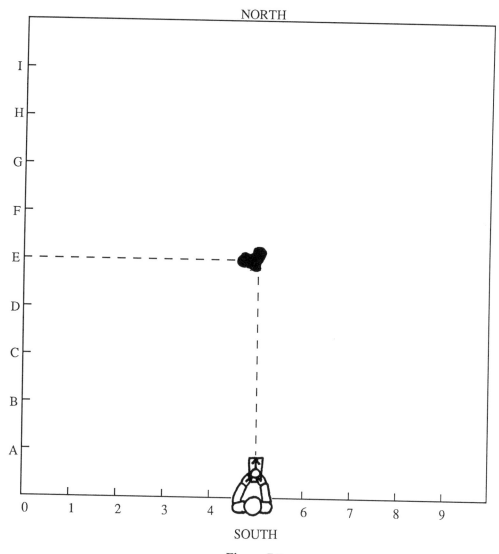

Figure 7.8

the line of the graph paper which represents the east boundary of the land.

2. In a similar way, chart the object's position along the east-west line. Have a student (surveyor) set a compass to north. Walk along the south boundary line until the arrow points to the object. The surveyor's position can be charted on the graph paper. (See Figure 7.8.)

3. With a ruler, draw a line from the mark on the south edge of the paper which is parallel to the north-south sides. Then draw a line from the mark on the east boundary straight west, parallel to the east-west edges. The object being mapped is then drawn at the intersection of these two lines. This object now has been transferred from the real land area onto a map representing it.

III. Mapping

A. Divide students into several small groups. Preferably each can have an adult leader. Provide each group with a sheet of graph paper, pencil, clipboard, a compass, and two ten-meter ropes marked off in meter intervals.

B. Students use these materials to make a map of a ten-meter square area of land which they have chosen.

1. Students line up one rope for a baseline, the other for a sideline. Both follow the cardinal directions.

2. They then transfer at least three objects within this area to a map on the graph paper.

3. Maps, with symbols or drawings representing actual objects, can be detailed as students wish.

4. Legends should be included at the bottom of their maps. They should show scales, directional symbols, and any symbols used on the maps.

IV. Conclusion

    A. Have the entire group reassemble to visit each of the areas. Surveyors give reports of their areas and present their maps.

    B. Discuss results and some difficulties involved in mapmaking.

V. Alternate Activities

    A. *Triangulation* — Place a ten-meter rope on the ground some distance from the object to be mapped. Draw the rope to scale at the bottom of a sheet of graph paper. Stand at one end of the rope and measure the angle of the object from the baseline with a compass. Stand at the other end and do the same. On the graph paper, draw angles at each end of the mapped rope and extend their sides until they intersect. The intersection shows the placement of the object. This method is used to plot an object in a body of water or in another inaccessible place.

    B. Other maps can be made of increasingly more difficult areas which contain additions/obstacles and contours.

Jigsaw

Jigsaw, a two-hour class designed as a culminating activity, incorporates other classes and activities of the outdoor education program. However, it also can be used in a school yard or park after an outdoor education experience; thus, it provides teachers with a way to "bring it all home."

In Jigsaw students and teachers conduct an intensive study of a plot of land fifty meters square. Four activities dealing with mapping, habitat study, weather, and human impact are conducted. Thus, pieces of the outdoor education curriculum are brought together.

Advantages of the class are numerous. First, it makes maximum use of teachers and other leaders so they are involved with small groups of students. Next, it provides students an opportunity to choose activities in which they wish to participate. Thus, students apply what they have already learned and delve deeper into an activity that interests them. Last, the class is done with any number of students, providing that enough leaders are available for the small groups.

Jigsaw is most effective as the last class of the outdoor education experience. First, an explanation of the class and a presentation of activities are given to the entire group. Students then must choose one activity they wish to pursue and join the leader of that activity. Each group has a sheet to explain its investigations with space for recording collected data. However, the group need not feel tied to the questions since students may come up with their own questions and ideas.

At the end, each group presents a short report of their findings to the entire group. After their own investigations and listening to others' reports, students get a real sense of the many facets of one small area of land. The instructor marks the fifty-meter square area with four corner stakes before class. The area should provide some variety in vegetation, elevations, and objects. The side should be lined up according to the cardinal directions so students can easily orient themselves.

Equipment:

1. Twenty-five or fifty-meter rope with yarn markers every five meters
2. One compass for each group
3. Paper and pencils, several for each group
4. Data sheets for each group
5. Several sheets of graph paper
6. Weather instruments (See Weather Mapping.)

I. Mapping (See Mapping page 98.)

    A. Decide which side of the fifty-meter square is the base and which corner is the starting point. For this explanation, the base is the south side and the southeast corner is the starting point.

    B. Create a conversion formula for changing paces into meters. Have a student count paces along a ten-meter section of rope. From this figure a formula, _____ number of paces per ten meters. A rough estimate is usually one and a half paces per meter, or three paces per two meters. This is a fairly easy conversion for students to figure. To convert paces to meters, divide the given paces by three and multiply by two. To convert meters to paces multiply by one and a half.

    C. Stand on the base at the starting point, face north, and then walk along the baseline, moving west. Keep moving until you face the first object to be mapped. Using a compass, make certain you are standing exactly in a north-south line with the object. The traveling arrow of the compass should point at the object.

    D. Write down the following information on the mapping chart. (See Figure 7.9.)

        1. Name of object
        2. Your position (measured in meters) along the baseline from the starting point (If you are between yarn markers, estimate your position.)
        3. The number of steps to the object, walking in a line straight north (Write this number on the chart, convert it to meters, and write that on the chart.)

    E. Continue this process until every major object and landmark in the area is mapped on the chart.

    F. Convert the chart to a map on graph paper. (See Mapping class.)

        1. Figure the number of meters per one-inch square on graph paper.
        2. Draw the baseline, sides, and top line of the fifty-meter-square area.
        3. Converting paces to meters when necessary, find the first object's position in the area and mark it with a dot on the chart.
        4. Continue this process until all the objects charted are mapped on the graph paper.
        5. Draw the objects on the graph paper, which is now an accurate map of the area.
        6. Make a legend in the corner, indicating the number of meters and/or paces in each one-inch square and any symbols used on the map. Be sure to include an arrow, indicating north, in your legend.

I. MAPPING CHART

Name of Object	Base Line Measurement	Paces	Paces in Meters	Other Information

Figure 7.9

II. HABITAT STUDY (See Habitat Hunt, page 16.)

A. Animal Signs:

Animal	Sign	Quantity	Location

B. Possible food sources:

C. Water sources, real and potential:

D. Ground cover:

E. Dominant type of vegetation (trees, shrubs, grassland, etc.):

F. Trees (if any):

    1. Dominant type (estimate):
    2. Estimate the ratio of dead trees to live trees. What does this tell you about the natural history and future development of the area?

G. Give several two-word descriptions that accurately describe your area. USE YOUR IMAGINATION.

H. List possible connections between the types of vegetation and the animals that inhabit the area.

I. Is there any connection between the ground cover and the canopy (tree cover)?

J. Make a food pyramid model using animals that inhabit your area. Try to include at least four levels in your example.

III. WEATHER (See Weather Mapping, page 90.)

A. Overall weather conditions (cloud cover, humidity, barometric pressure, etc.):

B. Measurements. (Chart four stations in somewhat varying places.)

	Description of Station		Station #1	Station #2	Station #3	Station #4
W I N D S P E E D	Wind Direction					
	Ground Level	Anenometer Reading				
		Actual Speed				
	3 Meters	Anenometer Reading				
		Actual Speed				
T E M P E R A T U R E	10 Centimeters Below Surface					
	Just Below Surface					
	Ground Level					
	1 Meter High					
	2 Meters High					
	Wind Chill Factor (Use 1 meter level temperature.)					
	Snow Depth					

C. What two stations had the largest differences in measurements? Why?

D. What specific measurements varied the most? The least? What factors might have caused these differences or similarities?

E. How does the physical environment affect weather? What specific factors (woods, terrain, etc.) determine weather conditions in your area?

F. How has weather affected the physical appearance of your area? How about the inhabitants (people, plants, animals)?

G. Predict the weather for the next twenty-four-hour period. (Optional)

H. What is "real" wind speed? How would you find out?

IV. HUMAN IMPACT

A. Look for human signs. Bring back any litter. When do you think these signs appeared on your area?

B. List ten items on your plot of ground that Native Americans of this area could have used to make a living.

C. Make up a Native American legend about your area using animals, terrain, type of vegetation, etc., as material. Use another sheet of paper, if necessary.

D. What did this area look like two hundred years ago? Draw a picture of how it might have appeared. Use the same procedure to describe how it looked fifty years ago.

E. How might this area look fifty years from now? Draw a picture. What factor determines how it will look then?

F. If you owned this land, what would you do with it? Why?

Winter Watch

Twig-o-mania (Twig Class)

Objectives:

1. To make students aware of similarities and differences in trees.
2. To help students use "process of elimination" questions to identify specific trees.
3. To help students understand the terms *alternate, opposite terminal bud,* and *bundle scar* in relation to trees.
4. To help students appreciate the uniqueness and importance.

Note: Students do not need to memorize the names of the trees they encounter. It is more important that they merely begin to notice the differences in trees and that they learn how to use a key to identify them.

Equipment:

1. Trees of various types included on the Tree Key marked with yarn
2. One copy of the tree key for each group, possibly laminated to a piece of cardboard to increase durability
3. Hard writing surfaces
4. Pencils
5. Four twigs from each of five different trees

This class, designed for winter, is written here for certain types of trees from a specific area. It can be adapted for other types of trees; and a tree key, using twigs, can be written for any types of trees. For spring and fall versions of this class, see Tree-mendous (Leaf Class).

I. Introduction

    A. Students meet in the Tree Encounter area. The instructor informs them that by the end of class, they can amaze their friends by identifying quite a number of trees with one magical sheet of paper.

    B. The importance of trees is discussed.

        1. What do trees do for us?
        2. What do they do for animals, mammals, and insects?
        3. How do they affect the land and weather?
        4. What would change if all the trees in this area were cut down?

    C. Discuss some differences and similarities seen in trees.

II. Twig Activities

    A. Students strive to place themselves in groups of four by matching their twigs with others from the same kind of tree. Groups then can share with the class what characteristics they noticed that allowed them to appropriately group.

    B. Each student finds another student with a different twig and shares one characteristic that is different.

III. Definitions for using the Tree Key

The instructor should explain the following terms which enable students to use the tree key found in Figure 8.1.

    A. *Opposite* and *alternate* refer to the branching of twigs and branches. Alternate means that the twigs grow out of the stem with no other twig opposite it on the stem. Opposite branches grow opposite one another out of opposite sides of the stem. (See Figures 8.2 and 8.3.)

    B. The *terminal bud* is the bud at the end of the twig from which the leaf grows. If the end bud is different from the side (lateral) buds and if the bud comes straight out the tip of the stem, the tree has terminal buds. If the end bud is the same as the side buds or it comes out the stem at an angle, it is not a terminal bud.

    C. *Bundle scares* are left from the old leaf and are found next to each side bud. Within these scars are vascular bundles that served as pipelines between the tree and leaf. How many are there? One, three, or many? A hand magnifying lens may help to decide. (See Figure 8.4.)

    D. A *strong odor* is noticed on some trees. Take a very small twig and break or crush it to release the strongest odor.

    E. A *tassel* resembles a hat tassel and is present only when there is no terminal bud.

    F. *More than one end bud* means more than one end bud is seen on the ends of some twigs.

Hug a Tree

If you wish to hug a tree without seeming rude, you must first get to know it. First select a cute tree and use the tree key to begin a casual conversation. There are two questions for each number on the key. To begin, ask the first question under 1. If the tree answers ''Yes,'' then proceed to question 2; if it says ''no,'' go to the next question 1. From that point continue as directed in the key. After getting to know the tree's name and personality, you are ready to give the tree its first hug.

Winter Twig Key for Twelve Common Trees

1. If the leaves are less than 1/4-inch thick (needles), go to 2.
1. If there are no living leaves on the tree, go to 4.

2. If the needles are less than 1/2-inch long and waxy, it is a cedar.
2. If the needles are more than 1/2-inch long, go to 3.

3. If needles come out from branch in cluster of two, it is a red pine.
3. If needles come out from branch in clusters of five, it is a white pine.

4. If the branches are opposite, go to 5.
4. If the branches are alternate, go to 6.

5. If the bark is rough, slightly orange in color, it is a box elder.
5. If the bark is smooth, slightly gray in color, it is a maple.

6. If there is no end bud, but a tassel, it is sumac.
6. If there is an end bud, go to 7.

7. If the end bud appears the same size and shape as the side bud, go to 8.
7. If the end buds are different from the side buds, go to 9.

8. If the branch has an odor and smooth twigs, it is a cherry.
8. If the branch has no odor and somewhat hairy twigs, it is an elm.

9. If a branch has a cluster of end buds, it is an oak.
9. If a branch has only one end bud, go to 10.

10. If a branch has a green end bud, it is a sassafrass.
10. If a branch has a non-green end bud, go to 11.

11. If the leaf scar has more than three bundle scars, it is a hickory.
11. If the leaf scar has only three bundle scars, it is an aspen.

Figure 8.1

IV. **Students can begin to use the Tree Key** (See Figure 8.1.)

A. This Tree Key identifies twelve common trees, although many more types exist. If desired, a more complex tree key can be devised to include them.
B. To identify a tree, start with question 1. If the question is true, proceed to question 2. If question 1 is not true, proceed to the second question 1. Continue this procedure until the name of the tree is discovered.
C. Have the class, as a group, pick a tree and use the key to demonstrate its use. Then split the class into three different groups and send each group to a different, previously marked area. In each area they should find six trees marked with yarn. Using the key, they should try to identify all six. Afterward, regroup and check the results.

V. **Conclusion**

A. Students find and stand by trees that remind them of themselves and choose one word to describe their similarities. They take turns sharing each word or simply shout them all at once on the count of three.
B. Before departing, students give their trees a kiss and a big hug!

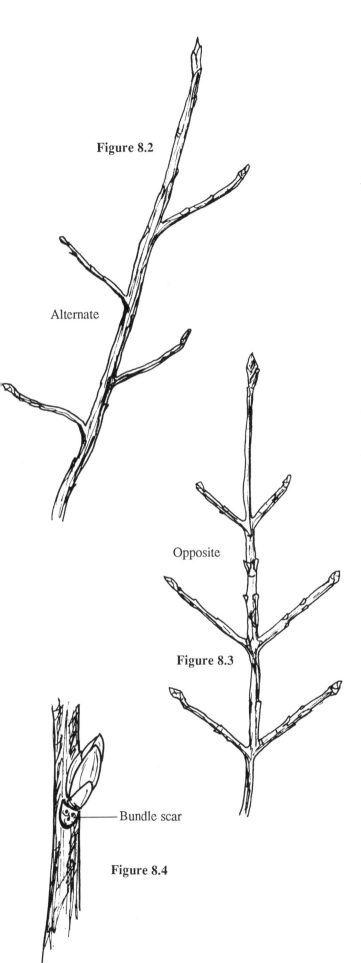

Figure 8.2

Alternate

Opposite

Figure 8.3

—Bundle scar

Figure 8.4

Winter Waterland

Objectives:

1. To help students gain an understanding of ice formation on a lake.
2. To teach students summer and winter differences in a lake.
3. To teach students that there is still life in the lake during winter and discover where that life is concentrated.

Equipment:

1. Golden Guide Books, *Pond Life* (See chapter bibliography.)
2. Charts showing most of the common swamp animals
3. Eight microscopes, including slides and blister slides (Single-power 50 X Blister Microscopes work well.)
4. Eye droppers, kitchen basters, petri dishes or small cups
5. Collecting jars or cups and buckets
6. Aquariums containing fish from a lake and a swamp aquarium
7. Two dip nets and a seining net for lake studies
8. Additional equipment for studies of the lake such as plankton nets, a Hach kit (to do PTT and other chemical tests), and extra dip nets, if desired
9. Ice spud to chip holes in the ice
10. Two plant hooks (long poles with bent pieces of wire tied on one end to gather submerged plants through the ice)
11. Fishing thermometer to record water temperature
12. A pond, swamp, or lake (Because the lake and swamp are frozen during the winter months, most of the animals are inactive and unavailable. Therefore, this class can be taught in the fall before the freeze and in the spring after the thaw.) (Also see the Aquatic Life class.)

I. Introductory Discussion

A. Water

1. What is water made of? It is composed of H_2O molecules.
2. What happens if you take a glass of hot water and a glass of cold water and pour one into the other? Before mixing together, the hot water rises to the top and the cold water drops to the bottom because cold water is heavier. Warm molecules move faster than cold molecules, so they spread apart more, making them lighter. This is also why hot air rises.

B. Ice Formation

1. What big change takes place in the lake as summer turns to winter? The water temperature drops.
2. As cool air hits the surface of the lake, it chills the top of the water and causes the surface water to sink to the bottom. Why? The cool surface water is heavier than the warmer deep water. Eventually, there is a shortage of oxygen.
3. As cool water drops, it pushes warm water up toward the top. Then, as the warm water gets

cooler, that also drops to the bottom. Keeping this in mind, where would ice form first? It seems that it would form at the bottom because cooler water is heavier. But where does ice first actually form? It forms at the top. Why? As water cools, molecules come closer together, making it heavier; however, that is only true down to the temperature of 39.2 degrees Fahrenheit (4 degrees centigrade). At that point, molecules mysteriously move apart again, making the cooler water lighter and causing it to remain at the top. Then at 32 degrees Fahrenheit (0 degrees centigrade), when water turns to ice, it is at the top.

4. Why does ice on a lake never get to be much more than a foot thick? The layer of snow and ice on the surface acts as a blanket, preventing the cold air from hitting the water; thus it keeps the water at a relatively constant temperature.

C. Winter Kill

1. What four things do all plants and animals need to survive? They require water, shelter, food, and oxygen. How do plants and animals that live in the lake during winter obtain each of those things?

   a. Water — They have continual access to water since they live in it.
   b. Shelter — They find shelter in plants, in the mud, under rocks, etc.
   c. Food — Plants can produce only food with the aid of sunlight. (See discussion in Habitat Study unit.) If the lake is covered with ice and snow during winter, how do plants receive sunlight? they cannot; as a result, they die, developing a food shortage by late winter.
   d. Oxygen — When the wind blows across the surface, creating movement in the water, oxygen is added. However, when ice prevents the wind from hitting the surface, the water becomes relatively stagnant.

3. As winter progresses, many plants and animals die as a result of the food and oxygen shortage. This result is known as *winter kill*.

II. Preparation for Collecting

A. Where would an animal, living in the pond during winter, go?

1. Bigger fish might go to deeper water where it is warmer.
2. Some animals might burrow into mud to hibernate.
3. Many small animals stay in shallow water among plants. Why? Plants offer both food and shelter; and the oxygen content is higher there.

B. Explain the procedure for collecting specimens.
C. Stress that the lake is dangerous and no one should wander away from the group. Although the ice in some places may be two feet thick, it may be five inches in other nearby spots. If an underwater spring keeps the water stirring, the ice tends to be thinner. Also, if a lot of decomposition occurs in a swampy area, heat is given off during the process, keeping the ice thin.

III. Procedure for Collecting

A. Using the ice spud (a long metal bar with a chisel-like end), chip a hole in the ice. A law states that a hole more than twelve inches across cannot be dug on a lake. Why? If it is bigger than a foot across, a person might fall completely through.
B. After the whole is dug, use plant hooks to scoop up any plants that may be growing underwater. Place the plants with some water into the large buckets and return to the classroom.

IV. Microscope Work

A. Find an animal in one of the jars that you want to observe. Suction the animal into a kitchen baster (large plastic dropper). Remember to squeeze the bubble of the baster before sticking it into the jar so it does not stir up the water.
B. Empty the baster into a shallow container.
C. Suction the specimen with a small eye dropper.
D. Empty the eye dropper onto a microscope slide.
E. Using your finger, wipe off some of the water.
F. Put the slide under the microscope and use the knobs to get the animal into sharp focus.
G. When you are finished looking, dip the slide into the water to let the animal go before it dries up and dies.

V. Alternate Activities

A. Go to a deep part of the lake and cut a hole. Using a weighted rope marked off in feet or meters, determine the lake's depth.
B. Compare the temperature of the lake at different depths using a fishing thermometer (which may be purchased at sporting goods store) attached to a graduated rope. Compare these with water temperatures taken close to shore in sand and mud.
C. Do similar experiments with ice thickness.
D. Plankton nets are fine-meshed, wire-rimmed, funnel-shaped nets directed into a small jar. The entire net is strung by a long cord. Lower the plankton net through a hole in deep water. As the net is pulled up, it captures many small-floating and free-swimming organisms (plankton), which can be brought back for observations under the microscope.

Winter Birds

Objectives:

1. To make students aware of the many differences in birds.
2. To help students understand the term *adaptation* and relate specific adaptations to birds' needs and environment.
3. To make students more aware of the methods used to classify and identify birds.
4. To help students recognize a few common winter birds.

Equipment:

1. A bird blind in a wooded area
2. Feeders in front of the bird blind viewing area
3. Seeds, corn, suet for feeders
4. Posted pictures or drawings of birds which may appear behind the bird blind.
5. Binoculars (optional)
6. As many types of study skins, feathers, beaks, feet, and wings of birds as possible
7. Copies of the Bird Discovery sheet and pencils (See Figure 8.5.)

I. What are birds? Why are they different from all other animals?

A. Birds have feathers.
B. Birds lay eggs.
C. They are warm-blooded and have a slightly higher body temperature than man.
D. Their wings are adapted for flight, and their feet are adapted for perching, walking, and swimming.
E. Their mouths are projecting beaks or bills with horny sheaths.

II. What is special about winter birds?

A. Why do some birds migrate and others do not? It is not the cold weather as much as the food supply that determines which ones migrate.
B. Which birds migrate? Which do not?

III. Adaptations, characteristics which help animals fit into their environment, are tools for survival.

A. Birds have many types of beaks. Name all the different kinds of foods birds eat. Beaks are adapted for these specific foods.

1. Long, narrow beaks work like tweezers to pick up insects.
2. Short, thick ones operate like nutcrackers to break open seeds.
3. Sieving, scooping, and straining beaks, which are like sieves, strain water plants and insects.
4. Hooked beaks work like knives to tear meat.
5. Nectar-sucking beaks work like straws.

B. Birds have feet and legs which are adapted to their needs and usually are determined by the types of food they eat.

1. Heavy claws, three toes in front and one in back, are used by predators because they need strong claws to catch, kill, and hold their prey.
2. Long, pointed claws, three in front and one in back, are used by seed eaters and scratchers.
3. Light claws, three in front and one in back, are used by songbirds for perching.
4. Long claws, two in front and two in back, are used by woodpeckers for climbing and balancing on the sides of trees.

Bird Discovery Sheet

1. Size: _____ Finger size _____ Hand size _____ Bigger than hand size

2. Colors: _____ _____ _____ _____

3. Type of Bill: _____ Short and thick _____ Long and narrow

4. Type of Feet: _____ Three toes in front, one in back _____ Two toes in front, two in back

5. Food: _____ Bird seed _____ Corn _____ Suet _____ Other

6. Actions: _____ Stays in one place to eat _____ Gets food and flies away

7. Where does the bird spend time? _____ on the ground _____ on branches _____ on trunk

8. Make up a name for the bird: _____

9. Other comments: _____

_____

_____

Figure 8.5

5. Webbed and lobed feet are used by water birds for swimming and walking.

C. Feathers are important adaptations. How do birds fly?

1. Wings have feathers and hollow bones and follow the shape of the body.
2. Feathers are used for warmth, flying silently, attracting mates, protecting nests, and for ruffling, which enlarges a bird's size to intimidate predators.

IV. Observation Skills

Pass around and discuss study skins, beaks, feet, feathers, and wings. Also discuss the proper way to handle them.

A. Examine the differences in various types.
B. Try to determine the food of each bird represented.

V. In the Bird Blind

A. Explain the use of binoculars before going to the blind.
B. Students should see numerous birds from the blind.

1. Students should be relatively quiet.
2. Movements, almost more than noise, frighten birds off; therefore, any movement outside the blind should be minimal.

C. Watch for differences in birds, which might be found in:

1. Behavior
2. Eating patterns
3. Types of foods
4. Feet and the way birds walk
5. Feathers and wings and the way birds fly
6. Beaks and how birds use them

D. Have a student hold seeds in one hand and stand outside. The student must be quiet and motionless. Often, if the student is still enough, a bird will light on his hand to eat.
E. A bird discovery sheet can be filled out by pairs of students working together. (See Figure 8.5.) They can fill out as many sheets on as many birds as they choose.
F. Mention that people feeding birds at the beginning of winter should continue through winter since birds become dependent on that food source and stay close when they normally might migrate.

VI. Other Activities

A. Make bird feeders or bluebird boxes.
B. Have students take apart a nest to see how it is constructed. Try to build a nest.
C. Give each pair of students a plastic bag and tweezers. One student must gather as much food

as possible and put it in the plastic bag "stomach." See which bird collects the most food in a given time.
D. Give students some tools (strainer, spoon, pair of pliers, scissors, tongs, and meat baster), representing birds' beaks. See which student can pick up the most beans or kernels of corn off the ground in a given time.

VII. Conclusion

A. Review characteristics of birds and the way their adaptations allow them to survive.
B. Discuss the relationship between birds and humans.

1. Birds are food.
2. Bird feathers are used in garments.
3. Birds are pets.
4. Birds consume harmful insects, rodents, and weed seeds.
5. Some birds damage crops.

Tracking

Objectives:

1. To help students improve their observation skills through observing and following tracks.
2. To help students realize that each animal has a distinct track and way of walking that is identifiable.
3. To give students a chance to track in a way similar to tracking a live animal.

Equipment:

1. Cards with drawings of various animal tracks found in the area
2. Fifty arrows drawn on pieces of cardboard and covered with plastic
3. Three pump spray bottles containing three distinctly different scents
4. Some sample plaster casts of tracks
5. Plaster of Paris
6. Mixing cups and spoons

I. Introductory Discussion

A. The history of tracking is as old as carnivorous man.
B. Native Americans were particularly good at tracking and shared their knowledge with the pioneers, mountain men, trappers and traders.

1. They tracked to find food.
2. They tracked to obtain pelts to trade for goods.
3. They tracked to learn from the animals.

II. Show track cards and plaster casts and discuss the types of animals that made them.

A. Have several students walk through sand, mud, or snow. Let them examine the differences in their tracks.

B. Have several students walk a short distance. Examine the differences in their strides: the heel or toe drag, the way they moved.

III. Discuss differences in tracks.

A. Size and shape identify the type of animal.
B. The depth of a track indicates an animal's weight.
C. The age of a track might be guessed based on how it has eroded, how the ground has dried since it was made, and where the track was made.
D. Tracks of the same animal may vary.

    1. Sand tracks are different from mud or snow tracks.
    2. Wind, rain, and sun alter tracks.
    3. The type of ground, soil, and ground cover affect tracks.

IV. Track Search

A. Split into groups and have students search for tracks.
B. Report findings to the entire class.
C. Examine some of the tracks and try to discover what type of animal made them.
D. Make plaster casts. Plaster should be mixed so it flows over the track. Add a little dirt or sand to help fill the track. Let the cast set for thirty minutes or so, perhaps longer.

V. Discuss other ways to track animals

A. Look for animal waste.
B. Find their homes, burrows, holes, nests, etc.
C. Look for their food and water sources. Go where the animal may go.
D. Watch for bits of fur, hair, or feathers.

VI. Tracking Activity

A. Three leaders each take cardboard arrows and one scent bottle, spraying the scent on their arrows and laying tracks. Arrows should be some distance apart, but not necessarily hidden.
B. The three trails can cross and wind back and forth over each other.
C. Students are divided into three groups and each group is given one of the three scents to smell. They then try to track their trail and find the leader at the end of it by smelling each of the arrows they find.
D. When snow covers the ground, cardboard arrows are not necessary because the scents can be colored with food coloring which shows up when the scents are sprayed on the snow. All three scents should be colored the same so students must smell stains to determine if they are part of their trail.

VII. Conclusion

A. Discuss some of the difficulties involved in trying to follow a track.

B. What are some difficulties involved in trying to follow only a smell?
C. Are people or other animals better trackers?
D. What does a person or an animal do when losing a track?

VIII. Other Activities

A. Time the deterioration of tracks by having students make, mark with date and time, and monitor tracks until they disappear.
B. Put out some food as bait in an untracked area of sand, mud, or snow. Check later for tracks.
C. Use a tame animal, which can be released and recaptured, and compare its activity with its tracks.
D. Split into two groups. Have one group make a trail using Native American signs with sticks and cairns. Send the second group to follow the trail and catch up with the Native Americans.
E. Have students draw tracks from books, photographs, models, or real tracks. This encourages them to be much more observant.

Skiing

Objectives:

1. Students have the opportunity to participate in a unique outdoor activity.
2. Students experience the winter landscape and, hopefully, gain enjoyment from it.
3. Students gain confidence, independence, and pride in succeeding at a new activity.

Equipment:

1. Skis, boots, and poles, enough variety in sizes to equip the necessary number of students (No-wax skis work best.)
2. Convenient storage area, racks, shelves to make outfitting of students easy
3. Tape or other means to mark all equipment by sizes

I. Introduction, History

A. Skies were used approximately 4,000 years ago. Old cave paintings show men hunting on skis.
B. Modern skiing originated in the Scandinavian countries, particularly in Norway.
C. Cross-country skiing also is known as Nordic skiing, while downhill also is called Alpine skin.

    1. Nordic skis are narrower and longer, allowing the skier to gain momentum from kicking off with one foot while gliding forward on the other, similar to ice skating. Also, the heel of the ski boot is not bound to the ski, so the skier can roll forward onto the toe while kicking off. Because of this, Nordic skis can be used for traveling uphill, downhill, or on flat land.
    2. Alpine skis are wider, shorter, and have a metal edge, enabling them to bite into snow or ice on the side of a slope during turns. The skier

gains momentum with the pull of gravity while going down a slope. The heel is bound to the ski, making skiing flat ground or up hill very difficult.

D. Cross-country skis, initially called skinny snow-shoes, were brought to the United States during the 1800s. Although a novelty, they were used in the west for traveling through deep snow. They provide a means of travel in the winter wilderness and actually are used like snowshoes. They support the body weight in deep snow like snowshoes, but are faster, particularly in downhill travel.

II. Parts of the Ski

A. The tip is the front, which is rounded for gliding, rather than digging into snow.

B. The tail is the back end.

C. The kick area is the part that touches the ground when body weight is on the ski, but is curved slightly off the ground when weight is removed. This area has the fish-scale pattern, mohair, or tacky wax that provides needed traction for the kick. The kick area allows the ski to glide forward, but holds it to snow when the skier kicks back to glide the other ski forward.

D. The cambor is the curve at the kick area. The ideal amount of cambor is determined by the weight and strength of the skier.

E. The binding is the metal part in the middle that holds the ski boot to the ski. In the past most cross-country skies have had a three-pin binding, which means there are three pins in a row at the top of the binding that align with three holes in the toe tip of the ski boot. With the pins in the holes, a clamp binds the shoe to the binding. In recent years other types of bindings, utilizing a simple clamp, have become available. Both types are effective. Bindings are on the skis at a slight slant, making right and left skis, which are indicated on the bindings.

F. The poles are used for added momentum and some balancing, but not for supporting the skier's weight.

G. The basket is the plastic ring at the pole's bottom that prevents the pole from going too far into snow. It also acts as a counter-balance for the skier while swinging the poles in stride.

III. Fitting Skis

A. Boots fit much like shoes: too loose, they cause blisters; too tight, they cut off circulation causing the feet to get cold.

B. Poles are measured to the armpit so the arm is parallel to the ground when outstretched.

C. To estimate the correct ski length, stand a ski on its tail. The ski is the correct length when its tip is equal to the height of the skier's wrist when the arm is stretched upward.

D. Cambor is not a major concern when outfitting groups with no-wax skis.

IV. Instruction

A. Students should be instructed in the basics of the kick-glide technique.

B. Instruction in the use of poles is fairly easy, since it is a very natural movement. After demonstrating how to hold the poles, it helps to mention that the poles follow the natural arm swing used when walking. When used correctly, the poles can add much to momentum.

C. Instruction should be given in turning.

1. The star turn, or step turn, is walking the skis around without crossing them. The tips can be kept together while the tails are walked around, or the tails can stay in position while the tips are walked around.

2. The kick turn is a faster way to turn and is done by kicking one ski into the air and flipping the tip over the tail so that one ski faces the opposite direction. Then the other ski is kicked around to line up with the first. This turn is a little more difficult; however, it is a fancy turn that students enjoy practicing.

D. Uphill techniques should be demonstrated and practiced.

1. Side-stepping is done by standing with skis perpendicular to the slope and side-stepping up the hill. This is best for ascending steep hills; however, it is a very slow process.

2. Herringbone is the design left in snow when a skier, with tails together and skis set on the inside edge, walks up a hill using poles from behind for extra support. This is a faster way to climb a hill.

3. When all else fails, a quick way to get up a hill, except in the deepest of snow, is to take the skis off and walk.

E. Downhill Techniques

1. The snowplow, a method of slowing down, involves putting the tips and knees closer together than the tails so skis are slanted onto the inside edge.

2. Place more weight on the ski on the inside of the turn to increase turning ability and decrease speed on a downhill.

3. The telemark turn, a method of turning used to control speed on steep slopes, involves bending the inside knee and rolling each ski toward its uphill edge.

4. When skiing too fast downhill and fearing loss of control, a controlled sit-down in the snow may be preferable to an uncontrolled fall.

V. Safety

A. Keep some distance between skiers, while preparing to ski and skiing. Be aware of poles and skis swinging around.

B. If falling on a hill, a skier should quickly roll off the trail to get untangled and back up.
C. If passing another, a skier should say, ''track left,'' to pass on the left and ''track right,'' to pass on the right.
D. Skiers coming down a hill have the right of way.
E. A skier should not start down a hill until the previous skier is safely at the bottom.

VI. Games

A. Tag
B. Follow-the-Leader
C. Red Light/Green Light
D. Formation Drills
E. Fox and Geese
F. Simon Says
G. Follow and Obstacle Course

BIBLIOGRAPHY

Brown, Tom, Jr. *The Tracker*. Berkley, CA: Berkley Publishing Company, 1978.

Murie, Olaus. *A Field Guide to Animal Tracks*. Boston, MA: Houghton Mifflin, 1954.

Stokes, Donald W. *A Guide to Nature in Winter*. Boston, MA: Little, Brown and Company, 1976.

BIBLIOGRAPHY

This limited selection of titles is only a partial listing of the resources that were gathered for preparing this title. *Nature's Classroom* has been an ongoing concern; a project over many years by many authors. Therefore many of the titles listed in this bibliography are possibly out of print or have been revised and republished since their use by the particular author of one of the activities. The titles appear here as they were used by the authors.

If you should have any information to clarify any of the following listings, please write to: Publications Department, American Camping Association, Bradford Woods, 5000 State Road 67 North, Martinsville, IN 46151-7902.

Alexander, Fichter. *Ecology, A Golden Guide.* Racine, WI: Western Publishing Company, 1973.

Angier, Bradford. *Feasting Free on Wild Edibles.* Harrisburg, PA: Stackpole Books, 1966.

Babcock, Harold L. *Turtles of North Eastern U.S.A.* NJ: Dover Publications.

Baker, Jim. *The Ways of the Warrior.* Worthington, OH: Hartland House, 1975.

Bealer, Alex. *Old Ways of Working with Wood.* Barre Publishing Co., 1972.

Berglund, Berndt and Bolsby, Clare F. *The Complete Outdoorsman's Guide to Edible Wild Plants.* NY: Charles Scribner's Sons, 1977.

Branson, Ann. *Soap.* New York: Workman Publishing Co., 1972.

Brown, Tom Jr. *The Tracker.* Berkley, CA: Berkley Publishing Company, 1978.

Contant, Roger. *A Field Guide to Reptiles and Amphibians.* Boston, MA: Houghton Miffling, 1978.

Curious Naturalist, The. NY: Massachusetts Audobon Society, September 1986.

Couchman, et al. *Mini Climates; Examining Your Environment.* Minneapolis, MN: Mine Publications, 1971

Couchman, et al. *Mapping Small Places; Examining Your Environment.* Minneapolis, MN: Winston Press, 1972.

Davis, Christopher, *North American Indian.* NY: Hamlyn Publisher., 1970.

Dickerson, Mary C. *The Frog Book.* NY: Dover Publications, 1969.

Edmunds, David R. *The Potawatomis, Keeper of the Fire.* Norman: University of Oklahoma Press, 1978.

Ekert, Allan W. *Tecumseh! A Play.* Toranto: Little, Brown and Company, 1974.

Gringhuis, Dirk. "Moccasin Tracks; A Saga of the Michigan Indian," *Museum of Michigan State University Education Bulletin.* No. 1, 1971.

Hunt, Ben W. *The Golden Book of Indian Craft and Lore.* Simon and Schuster, 1954.

Icenhower, Joseph B. *Tecumseh and the Indian Confederation 1911-1813.* NY: Franklin Watts Inc., 1975.

Jaeger, Ellsworth. *Nature Crafts.* NY: The Macmillan Co., 1950.

Jennings, Bertha W. *American Indian Society Cook Book.* Third printing, 1975.

Keller, Allan. "Pontiac's Conspiracy," *American History Illustrated.* Vol. XII, No. 2, May 1977, p. 4.

Klots, Elsie B. *The New Field Book of Freshwater Life.* NY: G. P. Putnam's Sons, 1966.

Kuhlman, Barbara. *Resident Outdoor Education Handbook.* Ridgewood Public Schools, 1975.

Lamb, Wendell E. and Shultz, Lawrence W. *Indian Lore.* Winona Lake, IN: Light and Life Press, 1964.

Lamb, Wendell E. and Shultz, Lawrence W. *More Indian Lore.* Winona Lake, IN: Light and Life Press.

Leasch, Alma. *Vegetable Dyeing.*

Maxwell, James A. ed. *America's Fascinating Indian Heritage.* Pleasantville, NY: Reader's Digest Association, Inc., 1978.

Miller, James A. and Holluns, Don. *Exploring Nature's Classroom.* Michigan Environment Education Association.

Murie, Olaus J. *A Field Guide to Animal Tracks.* Boston, MA: Houghton Mifflin, 1954.

Needham and Needham. *A Guide to the Study of Fresh-Water Biology.* San Francisco, CA: Holden Day, Inc., 1962.

Outdoor Activities. Environmental Science Center, 5400 Glenwood Ave., Golden Valley, MN 55422.

Parker, Arthur C. *The Indian How Book.* NY: Dover Publication, 1954.

Plants and Gardens. Two special printings: Vol. 29, No.2 and Vol. 29, No. 3.

Reid, Zim and Fichter. *Pond Life: A Guide to Common Plants and Animals of North American Ponds and Lakes.* NY: Golden Press, 1967.

Ritzenthaler, Robert E. and Rizenthaler, Pat. *The Woodland Indians of the Western Great Lakes.* NY: The Natural History Press, 1970.

Schneider, Richard C. *Crafts of the North American Indians.* Van Nostrand Reinhold Co., 1972.

Sloan, Eric. *A Museum of Early American Tools.* Funk, Inc., 1964.

Smith, Dr. Hobart M. *Snakes as Pets.* Fond du Lac, WI: All Pets Book, 1958.

Smith, Robert Leo. *Ecology and Field Biology.* NY: Harper & Row, 1966.

Stokes, Donald W. *A Guide to Nature in Winter.* Boston, MA: Little, Brown and Company, 1976.

Swan, Malcolm D. *Tips & Tricks in Outdoor Education.* Danville, IL: The Interstate Printers & Publishers, Inc., 1983.

Tunis, Edwin. *Colonial Craftsmen and the Beginnings of American Industry.* World Publishing, 1965.

Van Matre, Steve. *Sunship Earth.* Martinsville, IN: American Camping Association, 1980.

World Book Encyclopedia. Pioneer Life. Reprint. Field Enterprises Educational Corp., 1962.

INDEX